67

Advances in Biochemical Engineering/Biotechnology

Managing Editor: T. Scheper

Springer-Verlag Berlin Heidelberg GmbH

Influence of Stress on Cell Growth and Product Formation

With contributions by
H. J. Henzler, P. M. Kieran, G. Kretzmer,
P. F. MacLoughlin, D. M. Malone, W. Schumann,
P. A. Shamlou, S. S. Yim

Springer

Advances in Biochemical Engineering/Biotechnology reviews actual trends in modern biotechnology. Its aim is to cover all aspects of this interdisciplinary technology where knowledge, methods and expertise are required for chemistry, biochemistry, microbiology, genetics, chemical engineering and computer science. Special volumes are dedicated to selected topics which focus on new biotechnological products and new processes for their synthesis and purification. They give the state-of-the-art of a topic in a comprehensive way thu. being a valuable source for the next 3–5 years. It also discusses new discoveries and applications.

In general, special volumes are edited by well known guest editors. The managing editor and publisher will however always be pleased to receive suggestions and supplementary information. Manuscripts are accepted in English.

In references Advances in Biochemical Engineering/Biotechnology is abbreviated as Adv. Biochem. Engin./Biotechnol. as a journal.

ISSN 0724-6145
ISBN 978-3-642-08594-9 ISBN 978-3-540-47865-2 (eBook)
DOI 10.1007/978-3-540-47865-2

Library of Congress Catalog Card Number 72-152360

Originally published by Springer-Verlag Berlin Heidelberg New York in 2000
Softcover reprint of the hardcover 1st edition 2000

Typesetting: Fotosatz-Service Köhler GmbH, Würzburg
Cover: E. Kirchner, Heidelberg
Printed on acid-free paper SPIN: 10691471 02/3020 – 5 4 3 2 1 0

Preface

The local environment of the micro-organisms and cells in large industrial bioreactors varies over a broader range than in small laboratory reactors, spinner or shake flasks, because of the higher heterogeneity of the cultivation medium caused by the longer global mixing time. In addition the turbulence intensity and local shear forces are exacerbated with the increasing scale of the reactors. The strong fluctuation of these local values causes stress to the micro-organisms and cells.

In the last 20 years, several research groups have dealt with the investigation of the influence of stress on micro-organisms, animal and plant cells. The last review by J.C. Merchuk, published in 1991, dealt with shear effects [1]. However, recent molecular biological investigations indicate that the stress caused by fluid dynamic stress (shear and turbulence), extreme values of process variables (temperature, pH-value, concentrations of dissolved oxygen, glucose, phosphate, various salts), metabolically harmful substances (latic acide, ammonia, ethanol and various other solvents) and complex metabolic processes cause similar responses in the cells. These molecular biological fundamentals are treated in the first chapter entitled "Function and regulation of temperature inducible bacterial proteins on the cellular metabolism" by W. Schumann.

In this volume not all stress types are treated. Various aspects have been reviewed recently by various authors: e.g. "The effects of oxygen on recombinant protein expression" by Konz et al. [2]. The "Mechanisms by which bacterial cells respond to pH" was considered in a Symposium in 1999 [3] and solvent effects were reviewed by de Bont in the article "Solvent-tolerant bacteria in biocatalysis" [4]. Therefore, these aspects are not considered in this volume. Influence of fluid dynamical stresses on micro-organism, animal and plant cells are in center of interest in this volume. In chapter 2, H.-J. Henzler discusses the quantitative evaluation of fluid dynamical stresses in various type of reactors with different methods based on investigations performed on laboratory an pilot plant scales. S.S. Yim and A. Shamlou give a general review on the effects of fluid dynamical and mechanical stresses on micro-organisms and bio-polymers in chapter 3. G. Ketzmer describes the effects of shear stress on adherent cells in chapter 4. Finally, in chapter 5, P. Kieran considers the influence of stress on plant cells.

These five contributions provide an up-to-date report on stress effects is given.

November 1999 Karl Schügerl and Gerlinde Kretzmer

References

1. Merchuk JC (1991) Adv Biochem Eng Biotechnol 44:65
2. Konz JO, Kling J, Cooney CL (1998) Biotechnol Prog 14:393
3. Symposium on "Mechanisms by which bacterial cells respond to pH" held in London 2-4 June 1998. To be published in the book: Bacterial responses to pH in 1999 by John Wiley
4. De Bont JAM (1998) Tibtech 16:493

Contents

Function and Regulation of Temperature-Inducible Bacterial Proteins on the Cellular Metabolism

Wolfgang Schumann

Institute of Genetics, University of Bayreuth, D-95440 Bayreuth, Germany
E-mail: wolfgang.schumann@uni-bayreuth.de

Temperature is an important environmental factor which, when altered, requires adaptive responses from bacterial cells. While a sudden increase in the growth temperature induces a heat shock response, a decrease results in a cold shock response. Both responses involve a transient increase in a set of genes called heat and cold shock genes, respectively, and the transient enhanced synthesis of their proteins allows the stressed cells to adapt to the new situation. A sudden increase in the growth temperature results in the unfolding of proteins, and hydrophobic amino acid residues normally buried within the interior of the proteins become exposed on their surface. Via these hydrophobic residues which often form hydrophobic surfaces proteins can interact and form aggregates which may become life-threatening. Here, molecular chaperones bind to these exposed hydrophobic surfaces to prevent the formation of protein aggregates. Some chaperones, the foldases, allow refolding of these denatured proteins into their native conformation, while ATP-dependent proteases degrade these non-native proteins which fail to fold. Most chaperones and energy-dependent proteases are heat shock proteins, and their genes are either regulated by alternate sigma factors or by repressors. The cold shock response evokes two major threats to the cells, namely a drastic reduction in membrane fluidity and a transient complete stop of translation at least in *E. coli*. Membrane fluidity is restored by increasing the amount of unsaturated fatty acids and translation resumes after adaptation of the ribosomes to cold. Neither an alternative sigma factor nor a repressor seems to be involved in the regulation of the cold shock genes in *E. coli*, the only species studied so far in this respect.

Keywords. Heat shock, Cold shock, Genetic regulation, Chaperones, Proteases

Advances in Biochemical Engineering/
Biotechnology, Vol. 67
Managing Editor: Th. Scheper

1 Introduction

Microbial growth involves a complex series of integrated chemical reactions which are directly affected by external temperature. The most susceptible cell components are proteins, since subtle changes in their configuration may result in their inactivation. Cellular proteins, particularly regulatory proteins and key metabolic enzymes, require adjustments to cope with the temperature shifts that occur in natural or artificial environments and to allow balanced growth at the new temperature. The effects of temperature on bacterial growth can be expressed by the Arrhenius equation [1]. For most microorganisms, the slope of the Arrhenius plot is linear over a wide temperature range (about 30 °C). When temperature shifts are performed within this range, the growth rate usually changes rapidly, and the protein compositions of cells remain fairly constant. At temperatures above or below this range, the response is no longer linear and growth finally ceases. Rapid temperature shifts to the upper or lower limits have a pronounced effect on the physiology of bacterial cells, and the synthesis of specific sets of stress proteins is induced or enhanced [2].

In their natural habitat, bacteria are frequently submitted to large and rapid temperature changes and they can develop over a wide temperature range. This ability must be accompanied by adaptive changes in response to alterations of a large number of physical and chemical parameters, including solubility, reaction kinetics, membrane fluidity and protein conformation. Therefore, the biochemical effects following large temperature changes are likely to be complex, involving a number of interacting phenomena. For example, high and low temperatures may effect structural and functional changes in proteins to modify biological behavior by altering rates of enzyme activities, which may have important consequences for the integration of biochemical pathways. The conformational stability of proteins results from a complex balance of individual forces (hydrogen-bonding, electrostatic or hydrophobic interactions) which vary in different ways with temperature.

Bacteria possessing the ability to survive and multiply in a wide variety of environmental conditions are characterized by variability and adaptability. They are equipped with a large reservoir of genetic information encoding biochemical pathways which allow them to sense and to respond to changing environmental conditions by turning on genetic regulatory systems. This genetic repertoire has been recognized as an essential feature that enables bacteria to survive and successfully adapt to numerous stressful treatments, among them heat, cold, osmolarity, starvation, radiation and oxidative stress. Products of genes comprising these networks are involved in a broad range of cellular events, from receiving the initial signal to repairing the damage caused by stress. One example should illustrate the power of adaptation. *E. coli* contains seven *rrn* operons, and all these operons are required for *E. coli* cells to rapidly adapt to nutrient and temperature changes [3]. The time taken to adapt to a temperature shift from 28 to 42 °C increases with the decreasing number of intact operons. It has been suggested that the ability to adapt to changing environmental conditions provides the selective pressure to retain several operons.

Both the heat and cold shock response are universal and have been studied extensively. The major heat shock proteins (HSPs) are highly conserved. They are involved in the homeostatic adaptation of cells to harsh environmental conditions. Some act as molecular chaperones for protein folding, while others are involved in the processing of denatured polypeptides whose accumulation would be deleterious. The cold shock results in the transient induction of cold shock proteins (CSPs), which include a family of small acidic proteins carrying the cold shock domain. The CSPs appear to be involved in various cellular functions such as transcription, translation and DNA recombination.

2 The Heat Shock Response

When bacteria or almost any organism are exposed to high temperatures, the synthesis of a set of HSPs encoded by heat shock genes is rapidly and transiently induced; this reaction has been designated as the heat shock response. Later, it was found that most heat shock genes are not only induced by heat, but by many other stress regimens (Table 1). The stress factors have been classified into three groups: physicochemical factors, metabolically harmful substances and complex metabolic processes. Therefore, the heat shock is often called the stress shock to comply with these observations.

Table 1. Inducers of the *E. coli* heat shock genes

1. *Physicochemical factors*
   temperature-upshift
   pH
   osmolarity
   UV irradiation
2. *Metabolically harmful substances*
   ethanol
   antibiotics
   heavy metals
   DNA-damaging agents
3. *Complex metabolic processes*
   carbon and amino acid starvation
   viral infection
   oxidative stress

The heat or stress shock has been evolved in order to detect and deal with the presence of unfolded, misfolded, damaged or aggregated polypeptide chains collectively designated as non-native proteins in the following sections. To deal with the problem of protein unfolding/aggregation, two sets of universally conserved proteins have evolved, molecular chaperones and proteases. Molecular chaperone proteins act primarily by binding to the reactive hydrophobic surfaces of polypeptides to sequester them from the rest of the reactive surfaces present in their vicinity, thus effectively preventing aggregation and favoring the proper folding pathway (for details, see below). The chaperones act without covalently modifying their polypeptide substrates and without being part of the finished product. Non-native proteins which cannot be rescued by chaperones will be degraded by proteases. Work carried out during the last few years indicate that chaperones and proteases may work together in the degradation of non-native proteins, and one protein has been identified which acts either as a chaperone or as a protease depending on the growth temperature (see below). Therefore, chaperones and proteases are part of a regulatory network which maintains the quality control of cellular proteins.

3 Physiology of the Heat Shock Response and Thermotolerance

A stimulus (stress factor, stressor) from the environment effects a particular target or sensor which generates a signal which directly or indirectly affects the activity or synthesis of a regulator which, in turn, controls the output, usually a homeostatic response to the environmental change (Fig. 1). The signal may be an altered macromolecule or cell structure or a small molecule, either an organic metabolite or an inorganic ion, the level of which can be rapidly elevated or decreased. The regulator can be a repressor, an activator or programmer of RNA polymerase (alternative sigma factors). The products of the regulated operons then perform particular cellular functions to promote growth or survival in the new environment. Finally, modulators have to assure a return to the pre-stim-

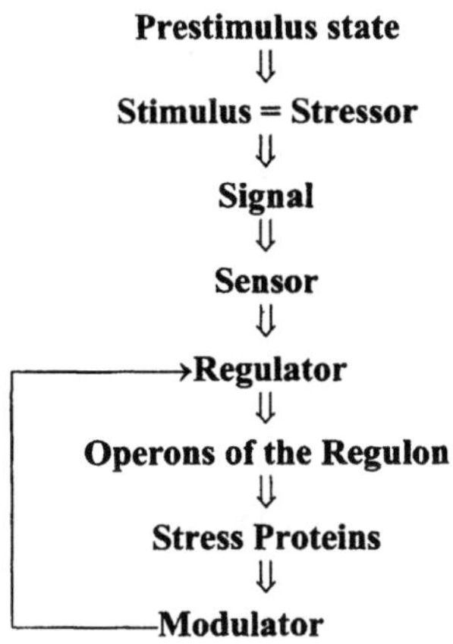

Fig. 1. Signal transduction pathway: Converts stress to a cellular response

ulus conditions of the cell or establish a new equilibrium consistent with growth. Return to the pre-stimulus state or the establishment of a new equilibrium consistent with growth is provided by a feedback loop. A familiar pattern is the inclusion of the gene for the regulatory protein (or proteins) as a regulated operon of the regulon itself.

It transpires that the heat shock genes of one single bacterial species are regulated by different mechanisms. Genes and operons controlled by one particular regulator are called regulons and, if there are at least two regulons in one species induced by the same stress factor, they form a stimulon.

When bacterial cells are shifted directly from a low growth temperature to a lethal temperature, the cells are rapidly killed. However, if the cells are first pre-adapted by growth at a non-lethal temperature for 30 min, the rate of killing upon a shift to lethal temperature is dramatically decreased; these cells have acquired thermotolerance. During the pre-adaptation phase a heat shock response is induced which leads to increased synthesis of heat shock proteins.

4
Major Heat Shock Proteins and Their Function

The known major heat shock proteins encoded by the *E. coli* chromosome are summarized in Table 2. Most of them are either molecular chaperones or energy-dependent proteases. Both types of enzymes together serve to maintain the quality control of cellular proteins and have as their substrates a variety of misfolded and partially folded proteins that arise from slow rates of folding or assembly, chemical or thermal stress, intrinsic structural instability and biosynthetic errors leading to non-native proteins. The primary function of classical chaperones is two-fold: First, to modulate protein-folding pathways, thereby preventing misfolding and aggregation, and second, to promote refolding and proper assembly. Energy-dependent proteases recognize and select their protein targets in an environment of other proteins that should not be degraded. It

Table 2. Important *E. coli* heat shock genes, proteins and their function

Gene	Protein	Function
ATP-dependent molecular chaperones: foldases		
dnaK	DnaK	chaperone
dnaJ	DnaJ	cochaperone of DnaK; stimulates the ATPase activity of DnaK
grpE	GrpE	nucleotide exchange factor of DnaK
groEL	GroEL	chaperone
groES	GroEL	co-chaperone of GroEL
clpA	ClpA	chaperone
clpX	ClpX	chaperone
clpY	ClpY	chaperone?
degP	DegP	chaperone at low and protease at high temperatures
Molecular chaperones: holdases		
hslO	Hsp33	chaperone with a redox switch
htpG	HtpG	chaperone of unknown function
ibpA	IbpA	chaperone found in association with some inclusion bodies
ibpB	IbpB	chaperone found in association with some inclusion bodies
ATP-dependent proteases		
lon	Lon	
clpP	ClpP	acts together with either ClpA or ClpX
ftsH	FtsH	protease and probably also chaperone
clpQ	ClpQ	acts together with ClpY
Other important heat shock proteins		
clpB	ClpB	renaturation or proteolysis of aggregated proteins
yrfH	Hsp15	binds to DNA and RNA
htpY	HtpY	regulation of the heat shock response; exact function unknown
rpoD	σ^{70}	vegetative sigma factor
ftsJ	FtsJ	methylates rRNA
lysU	LysU	lysyl-tRNA synthetase II
ppiD	PPiD	peptidyl-prolyl isomerase
fkpA	FkpA	peptidyl-prolyl isomerase
pspABCD	PspABCD	stationary phase survival and protein transport

seems reasonable to expect that protein unfolding is a necessary prerequisite for attack by a protease. Recent work has demonstrated that some ATP-dependent proteases at least have intrinsic chaperone activity, suggesting that the initial steps in energy-dependent protein degradation may be similar to those of chaperone-dependent protein folding. In addition, very interestingly, most, but not all, chaperones and many proteases are heat shock proteins underlining their importance in protein quality control under stressful conditions.

4.1 Molecular Chaperones

Molecular chaperones are able to temporarily stabilize unfolded or partially folded proteins and prevent inappropriate inter- and intramolecular interactions. They reduce the free concentration of aggregation-sensitive folding intermediates and effectively prevent aggregation processes both *in vivo* and *in vitro* (for a recent review, see [4]). Molecular chaperones are divided in two major classes according to their dependency on energy. Both ATP-dependent and ATP-independent chaperones seem to act in concert to guarantee survival of the cell under stress conditions, and only one chaperone is essential under all growth conditions (GroEL/ES). While the ATP-dependent chaperones influence the yield and kinetic of folding, the ATP-independent ones suppress aggregation of misfolded proteins. To distinguish between these two activities, the terms "foldases" and "holdases" have been coined [5].

Heat-inducible chaperone systems in the cytoplasm of *E. coli* include the DnaK/DnaJ/GrpE system, the GroEL/GroES complex, ClpA, ClpB, ClpX, ClpY, HtpG, Hsp33 and small heat shock proteins. The most abundant and well-studied members are the DnaK and the GroEL systems (see recent reviews by Rassow et al. [6], Bukau and Horwich [4], and Richardson et al. [7]). The DnaK protein consists of a highly conserved N-terminal ATPase domain of 44 kDa and a C-terminal region of 25 kDa, divided into a conserved substrate binding domain of 15 kDa and a less-conserved immediate C-terminal domain of 10 kDa. The ATPase domain consists of two large, globular subdomains, separated by a deep central cleft and connected by two crossed helices. Both subdomains and the connecting helices contribute to the binding pocket for nucleotide and the required Mg^{2+} and K^{+} ions at the bottom of the cleft. The first part of the substrate-binding domain is folded into a compact sandwich containing two four-stranded antiparallel sheets; the second part consists of helices and is more extended. Substrate proteins bind to the sandwich in an extended structure, and the helical part serves as a lid on the peptide-binding unit without direct contact with the substrate where hydrophobic polypeptides of at least seven residues are optimal partners.

The central residue of the heptamer is of particular importance for high-affinity binding as it protrudes into a relatively large hydrophobic pocket within the sandwich. The helical lid can adopt two different states, open and closed, and the mechanism which regulates the transition between both stages in the context of hydrolysis of ATP is under investigation. The DnaJ protein contains a highly conserved domain of 80 amino acid residues, the J-domain, often located near the N-terminus, which is essential for the stimulation of the DnaK ATPase activity. The NMR-based structure of this domain describes a scaffolding of four helices, at the tip of which lies the conserved tripeptide His-Pro-Asp in a solvent-accessible loop region; substitutions within this tripeptide abolish the interaction with DnaK. The J-domain is followed by a glycine/phenylalanine-rich region, a cysteine-rich central domain and a C-terminal stretch. The central region contains two zinc atoms, each coordinated to four cysteines as in the zinc fingers of DNA-binding proteins. This region and part of the C-terminal domain participate in binding unfolded polypeptides. The nucleotide exchange factor GrpE consists of 34 N-terminal amino acid residues of unknown function, followed by the central dimerization domain and the C-teminal region where the latter two are involved in DnaK binding. GrpE triggers nucleotide exchange by a contact through one GrpE subunit that opens the nucleotide binding cleft of DnaK.

All three components of the DnaK/DnaJ/GrpE system, often called the DnaK chaperone machine, act together in a chaperone cycle to allow refolding of non-native substrate proteins. The reaction cycle starts with the association of DnaJ with a substrate, followed by transfer of the substrate to the peptide-binding site of the ATP form of DnaK which is present in the open state. Binding of the substrate together with DnaJ stimulates hydrolysis of ATP in the ATPase domain of DnaK; the chemical energy of ATP is used to perform mechanical work, the opening and closing of the substrate binding pocket. As a consequence, the peptide-binding domain adopts the closed state. Upon interaction with GrpE, ADP is released from the ATPase domain. After binding of a new molecule of ATP, the bound substrate is released from the open state, and either the same or a new substrate molecule can bind. Substrate molecules can undergo several cycles by binding and release until they have adopted their native three-dimensional structure. Alternatively, partially refolded polypeptide chains can bind to the GroEL chaperone machine (see below).

The GroEL/GroES system consists of two different components, the GroEL and the GroES proteins. GroEL is composed of 14 subunits of 58 kDa each, arranged in two head-to-head rings in whose central non-connected cavities various substrate proteins can be transiently sequestered from the medium and allowed to fold into their native conformation. GroES is composed of seven subunits of 10.5 kDa each that form a dome capping GroEL's central cavity in the presence of nucleotides. Each GroEL subunit consists of three different domains: an equatorial domain which is responsible for binding and hydrolysis of ATP and for all the inter-ring and most of the intra-ring interactions; an apical domain, which contains all the residues involved in substrate and GroES binding; and an intermediate domain, which connects the apical and equatorial domains and acts as a hinge for the movements of the apical domain in response to signals generated in the equatorial domain. Upon ATP binding to the equatorial domains, GroES can bind to one or both GroEL rings leading to an asymmetric or symmetric complex, respectively. Binding of ATP to the equatorial domain induces an upward and outward movement of the apical domain that is further increased upon GroES binding. The mechanism of GroEL/GroES-assisted protein folding starts with an asymmetric complex where the GroES dome is bound to the lower GroEL ring, and where all subunits of the lower ring have bound ATP. The non-native protein will enter the cavity of the open upper ring and interact with a large patch of hydrophobic residues placed in the apical domain. This binding is first followed by hydrolysis of ATP in the lower ring and then by rapid binding of ATP and GroES to the upper ring. This binding will lead to the release of the non-native protein from the apical domain into the interior of the cavity and a large conformational change within the upper ring which will double the cavity size. Once released the non-native protein starts to fold into its native conformation. ATP hydrolysis in the upper ring destabilizes the high-affinity GroEL/GroES complex. GroES release is stimulated by ATP binding to the lower ring. The substrate protein is now released from the cavity of the upper ring and, if not folded properly, it can rebind to the same or a different GroEL molecule. The whole cycle of binding and release is repeated as many times as required for the substrate to attain its native structure.

The GroEL chaperone captures non-native proteins by means of a ring of hydrophobic residues that line the entrance to the central cavity of its heptameric ring. When GroEL binds ATP and GroES, a massive structure change doubles its hydrophobic binding surface. The substrate protein is transiently encapsulated in the central cavity under the GroES lid. The manner in which the GroEL structure change promotes protein folding remains to be demonstrated.

Two models, not mutually exclusive, are under consideration. The Anfinsen cage model is based on the view that protein folding is limited by intermolecular reactions that produce aggregation. The model proposes that the GroEL cavity provides a sequestered microenvironment where folding to the native state can proceed while the substrate protein is protected from aggregation. However, the substrate protein is ejected from the cavity with each round of ATP hydrolysis whether it has reached the native state or not. The iterative annealing model is based on the view that the rate-limiting step in slow protein folding is the intramolecular reorganization of misfolded and trapped protein segments, dependent on some degree of protein unfolding. This model proposes that ATP hydrolysis is coupled to a forceful unfolding of the misfolded substrate protein and its release, either into the protected central cavity or to the exterior, so that the misfolding is relieved and forward folding can resume. Incompletely folded proteins undergo further iterations, in the biological equivalent of optimization through annealing, until they reach the native state.

Clp proteins have roles in many cellular processes including protein reactivation, protein degradation, DNA replication, regulation of gene expression and thermotolerance. The ATP-dependent ClpA and ClpX proteins by themselves act as chaperones, but they can also associate with ClpP, the proteolytic subunit (see below). The genes have been named *clp* because they code for *c*aseino*l*ytic *p*roteases. Both proteins are ATPases and form a six-membered ring. For ClpA, it has been shown in *in vitro* experiments that it activates the latent DNA-binding activity of the plasmid P1 replication initiator protein RepA by remodeling inactive dimers into active monomers with the help of ATP [8]. ClpA also prevents irreversible heat inactivation of luciferase *in vitro* [9]. Similarly, it was also shown that purified ClpX protein can act as a bona fide chaperone since it protects phage λO protein from heat-induced aggregation and is able to dissociate heat-induced λO aggregates with the help of ATP and activates DNA binding by the TrfA replicator initiator of plasmid RK2 by converting inactive dimers to active monomers [10, 11]. Furthermore, ClpX stimulates replication of bacteriophage Mu by promoting the dissociation of stable complexes of MuA tetramers and Mu DNA, allowing initiation of phage DNA replication [12, 13].

Another important protein of the Clp family is ClpB which possesses ATPase activity. In a *clpB* mutation, 45% of the denatured and aggregated protein arising transiently after the transfer of an *E. coli* culture from 30 to 45 °C is stabilized [14]. ClpB seems to play an important role in the renaturation or proteolysis of the aggregated proteins, but the mechanism of action of ClpB is not yet known. One can suppose that it might participate in the resolubilization of aggregated proteins.

Very recently, a novel class of chaperone activity has been described in *E. coli* [15]. This heat shock protein, termed Hsp33, is distinguished from all the other known molecular chaperones by its mode of functional regulation. Hsp33 is a cytoplasmic protein whose activity is regulated at the post-translational level by the environment. While the inactive, reduced form of Hsp33 has zinc coordinated by four conserved cysteine residues, oxidizing conditions cause zinc to be released from the cysteines, disulfide bonds are formed and the chaperone becomes active. Therefore, Hsp33 seems to be a chaperone with an on-off mode of activity that uses reactive disulfide bonds as a molecular switch.

It is generally believed that energy-independent chaperones are unable to allow refolding of non-native proteins. They bind to hydrophobic surfaces of denatured proteins to prevent aggregation and pass them to energy-dependent chaperones such as the DnaK or the GroE chaperone machines.

IbpA and IbpB have been identified as *E. coli* proteins that associate with some inclusion bodies formed after high-level expression of heterologous proteins [16, 17]. The genes *ibpA* and *ibpB* are organized in a common operon and encode proteins that share 52% identical amino acid residues to each other and to a group of heterogeneous proteins called small HSPs (molecular mass: 15–30 kDa) that share conservation of sequence in certain localized regions [16, 18]. Several members of this group have been shown to function as chaperones (holdases).

The HtpG protein shares 40% amino acid identity with the eukaryotic members of the HSP90 family [19]. Purified HtpG acts as an ATP-independent chaperone *in vitro* [20, 21]. It recognizes and transiently binds non-native folding intermediates, reducing their free concentration in solution and thus prevents aggregation. These characteristics qualify HtpG as a holdase. The *htpG* gene is non-essential; however, *E. coli* and *B. subtilis* strains with *htpG* null mutations exhibit a minor growth disadvantage at high temperatures [19, 22].

4.2
Energy-Dependent Proteases

Five important proteases belong to the group of heat shock proteins: Lon, ClpP, FtsH, ClpQ and DegP. The Lon protease is an oligomeric enzyme composed of identical 87 kDa subunits. Each Lon monomer carries a protease active site serine and a single consensus ATP-binding site [23–25]. Lon has ATPase activity that is activated by protein binding. While ATP hydrolysis is essential for degradation of proteins, ATP binding alone is sufficient to activate cleavage of small nonstructured proteins and oligopeptides. Protein substrates bind Lon at two sites, the proteolytic active site and an "allosteric site" that probably lies within the ATPase domain. Binding of potential substrates to the latter site activates ATPase activity. Lon resides in the cytoplasm and participates in the degradation of both specific protein targets (e.g., in *E. coli* the λN protein, the SulA cell division regulator and the positive regulator of capsule synthesis, RscA) and abnormally folded proteins. Many temperature-sensitive mutations are suppressed by *lon* mutations, implying that the defect in activity of these mutants at high temperature is due at least in part to proteolysis of the mutant protein. Proteins containing abnormal amino acids are degraded rapidly, and their proteolysis is slowed down significantly in *lon* mutants [26]. Of the energy-dependent proteases, Lon appears to be one of the best conserved throughout the bacterial world.

Some of the Clp ATPases, including *E. coli* ClpA, ClpX and ClpY, are regulatory components of two-component ATP-dependent proteases. Both ClpA and ClpX can activate degradation by ClpP, a second important heat-inducible protease. ClpP itself exists as a stable tetradecamer, composed of two stacked seven-membered rings, which is able to degrade small peptides *in vitro*. To gain protease activity, it has to assemble with either of two partners, ClpA or ClpX. Complexes of ClpP and ClpA contain a tetradecamer of ClpP flanked at one of both ends by a hexamer of ClpA [27]. Similar structures are seen with ClpXP and ClpYQ complexes. Thus, the Clp ATPase components are situated in an ideal position to regulate the entry of specific substrates into the aqueous core of ClpP.

In many cases the proteins that are substrates for Clp ATPase chaperone activity are also degraded by the corresponding Clp protease As already mentioned, ClpA activates RepA, and ClpAP degrades RepA [9]. Similarly, ClpX dissembles MuA-DNA complexes and disaggregates λO; ClpXP degrades MuA and λO [28, 29]. Therefore, the bipartite ClpAP and ClpXP proteases are composed of two types of polypeptides, the ClpA/X ATPase and the ClpP peptidase. Both reactions require ATP hydrolysis. These observations, among others, led to the idea that chaperones and proteases might act together where the chaperone is involved in unfolding the protein substrate and then presents it to the protease; these protease-associated chaperone activities have been designated charonins [30].

ClpYQ forms an additional two-component energy-dependent protease. While the *clpY* gene encodes a protein bearing extensive homology to ClpX ATPase, *clpQ* codes for a protein homologous to the proteasome subunit of eukaryotic ATP-dependent proteases and has a proteolytic active site. ClpY forms one ring of six or seven subunits and contains one ATP-binding site per subunit, and ClpQ is composed of two rings of six subunits [31–33]. It is not known whether ClpY alone has chaperone activity as has been shown for ClpA and ClpX.

The FtsH protease is anchored to the cytoplasmic membrane through two transmembrane domains where the short N-terminus and the long C-terminus are exposed to the cytoplasm (for a recent review, see [34]). The C-terminal part contains ATPase activity and Zn^{2+} metalloprotease active sites [35, 36]. In *E. coli*, FtsH forms a complex with a pair of periplasmically exposed membrane proteins, HflK and HflC, and is involved in the proteolytic degradation of some unstable proteins that include both soluble regulatory proteins such as the heat shock sigma factor σ^{32} and phage λ protein CII, a transcriptional activator. In addition, it can degrade membrane proteins without partners, e.g. SecY which forms the translocon together with SecE and SecG and the *a* subunit of the F_0 complex of the H^+-ATPase [37, 38]. The protease activity of FtsH is modulated by the HflKC complex and by another protein termed YccA; the latter protein is also an integral membrane protein, and can transiently associate with FtsH and serve as a substrate [39]. That the FtsH protein might also express chaperone activity in addition to its protease activity has not yet been strictly proven but the theory has been strengthened by several observations including the finding that it associates with denatured alkaline phosphatase without degrading it [40]. Therefore, the major task of FtsH might be to maintain quality control of some cytoplasmic and membrane proteins.

The last example of a heat-inducible protease with a particular property is the DegP protease (also called HtrA) which resides in the periplasm and belongs to the sigma-E regulon. DegP is a member of a large family of related serine proteases and His-105, Asp-135 and Ser-210 are postulated to be the catalytic triad residues [41]. This widely conserved heat shock protein has both chaperone and proteolytic activities [42]. At low temperatures, the chaperone function dominates and DegP is able to stimulate refolding of chemically denatured substrates. In contrast, its proteolytic activity is almost exclusively present at elevated temperatures. It seems plausible that at low temperatures folding defects are more easily repaired than at high temperatures where degradation of non-native proteins dominates. What is known about the mechanism of the temperature-dependent switch from chaperone to protease and *vice versa*? Here, the active site Ser-210 seems to play an important role. If this residue is changed to an alanine, DegP loses its protease activity, but retains its chaperone activity. Furthermore, the chaperone activity is located in the protease domain. The simplest model suggests that the protease domain of DegP can always bind non-native proteins. Whether the substrate enters cycles of binding and release or is proteolytically degraded can be controlled by the accessibility of the active site Ser-210. At low temperatures, the Ser-210 can be buried and become properly exposed after a conformational change induced by the shift from low to high temperatures [42].

5
Regulation of the Heat Shock Response

Regulation of the heat shock response has been studied most thoroughly in both *E. coli* and *B. subtilis* and, to some extent, in other bacterial species including *Bradyrhizobium japonicum* and *Streptomyces*. It turns out that (i) all known heat shock genes are induced at the level of transcription; (ii) they are either under the positive control of an alternate sigma factor or under negative control of a transcriptional repressor, (iii) the heat shock response is transient; and (iv) the heat shock genes present within one bacterial species form several regulons regulated by different mechanisms. Table 3 gives a summary of all known bacterial heat shock regulators.

In *E. coli*, the heat shock response is induced very rapidly, and an enhanced synthesis of heat shock proteins can already be seen after 1 min. This means that any kind of regulation cannot involve transcription and translation of a regulator gene. In the case of positive regulation it follows that the regulator must already be present before heat shock and kept inactive. In those cases studied, the positive regulator, always an alternative sigma factor, is kept in an inactive state by an anti-sigma factor. Upon a upward shift in temperature, it is released immediately from its anti-sigma factor, binds to the RNA polymerase (RNAP) core enzyme and thereby ensures transcription initiation. In the case of negative regulation by a repressor, this protein has to be inactivated within a very short time. Furthermore, each regulation mechanism has to be fine-tuned in such a way that induction of the heat shock response is correlated with the extent of stressor. While a modest heat shock response should result in only a slight induction of the heat shock genes, a severe heat shock should be followed by a dramatic increase in the expression of the heat shock genes.

5.1 Regulation by Alternate Sigma Factors

In *E. coli*, and most probably all enteric bacteria, all heat shock genes are controlled only by alternate sigma factors. In *B. subtilis* and its relatives, a large group of about 100 heat shock genes are also under the positive control of an alternate sigma factor. These sigma factors are present at all times and, therefore, their activity has to be regulated in accordance with the requirements of the cell. This occurs, in all but one case, by interaction with an anti-sigma factor.

5.1.1 *Sigma-32*

The first alternate sigma factor which plays a crucial role in regulating heat shock genes has been identified in *E. coli* and called σ^{32}; this sigma factor is encoded by the *rpoH* gene [43] and complexes with the RNAP core (E) to constitute $E\sigma^{32}$. $E\sigma^{32}$ recognizes the classical heat shock promoters located upstream of all the heat shock genes belonging to the sigma-32 regulon such as *dnaK*, *groEL* and *lon* (see Table 3; for recent reviews, see [44, 45]).

Table 3. Known bacterial heat shock regulators

Regulator	Occurrence in bacteria
Alternate sigma factors	
σ^{32}	*E. coli*, enterobacteria and several other Gram-negative species
σ^{E} (σ^{24})	*E. coli*, *Pseudomonas aeruginosa* and many other Gram-negative species
σ^{54}	*E. coli*
σ^{B}	*B. subtilis*, *Staphylococcus aureus*
Transcriptional repressors/operators	
HrcA/CIRCE (*c*ontroling *i*nverted *r*epeat of *c*haperone *e*xpression)	
	detected in *B. subtilis* and occurrence in bacteria most widespread repressor throughout the bacterial kingdom
CtsR/CtsR box (*c*lass *t*hree *s*tress gene *r*epressor)	
	detected in *B. subtilis* and present in many low G+C Gram-positive bacteria
HspR/HAIR (*H*spR *a*ssociated *i*nverted *r*epeat)	
	detected in *Streptomyces* and present in several bacterial species
ROSE (*r*epression *o*f heat *s*hock gene *e*xpression)	
	detected in *Bradyrhizobium japonicum*

The *rpoH* gene is expressed from four different promoters. Under normal physiological conditions, the P1 promoter is responsible for most of the *rpoH* transcription, while P2 and P4 promoters contribute varying minor amounts. The P3 promoter is under the control of the $E\sigma^E$ holoenzyme (see below) and becomes induced at temperatures above 45 °C. The *rpoH* gene is expressed at all temperatures, and after a heat shock its transcription is increased by a factor of 1.5 only, but there is a large transient increase in intracellular σ^{32} levels. Two factors contribute significantly to this increase: an enhanced rate of translation of the *rpoH* mRNA, and a transient stabilization in the half-life of σ^{32}.

Two important *cis*-acting regulatory regions have been identified as being involved in the regulation of *rpoH* mRNA translation. The first region, designated region A, is located near the 5′ end of the transcript and plays a positive role, since its deletion results in a 15-fold decrease in the rate of *rpoH* translation [46]. The second region B lies about 200 nucleotides downstream of the 5′ end. It behaves as a negative element, since its removal results in a substantial increase in the rate of *rpoH* mRNA translation. Certain nucleotides of regions A and B take place in base pairing resulting in the development of a secondary structure involving both the Shine-Dalgarno and the start codon thereby limiting access of the RNA polymerase holoenzyme. This secondary structure is transiently removed after a heat shock, allowing enhanced translation of the transcript. This observation led Yura and co-workers to suggest that the *rpoH* mRNA acts as a cellular thermometer [47]. A third *cis*-acting region, region C, is located in the center of σ^{32} and is involved in the half-life of this sigma factor [48]. Within region C, a segment of nine amino acids is almost entirely conserved within σ^{32} homologs and was therefore termed the RpoH box [49]. A high-affinity DnaK binding site exists within the RpoH box, and a second binding site was found close to the RpoH box at the periphery of region C making this region a prime candidate for a regulatory site which allows binding of DnaK [50].

The σ^{32} polypeptide has been shown to be one of the most unstable proteins of *E. coli*, its half-life being 45–60 s between 30 and 42 °C, but substantially longer at 22 °C [51, 52]. Following a shift from 30 to 42 °C, σ^{32} is dramatically stabilized for about 5 min, after which its extreme instability is restored. Furthermore, in a *dnaK*, *dnaJ* or *grpE* mutant background, and in the absence of FtsH, the half-life of σ^{32} is substantially increased at all temperatures [52–54]. These results strongly suggest that the DnaK chaperone machine indirectly and FtsH directly participate in σ^{32} degradation, and it has been shown that FtsH can cleave purified σ^{32} proteolytically [36].

How is the the heat shock response involving σ^{32} induced in *E. coli*? As stated above, the products of the four σ^{32}-dependent heat shock genes, namely *dnaK*, *dnaJ*, *grpE* and *ftsH*, negatively modulate the rate of their own expression. All the accumulated data are consistent with the interpretation that the intracellular levels of non-native proteins control the extent of the heat shock gene expression. These unfolded proteins will be recognized by the chaperones, and titration of the DnaK chaperone machine by these non-native polypeptides will increase the stability of σ^{32}. Furthermore, enhanced translation of its mRNA will increase the amount of σ^{32}. The more non-native proteins that have been removed from the cytoplasm, the more complexes of the DnaK chaperone team are available to sequester σ^{32}. It is still an open question as to how translation could be inhibited so that the *rpoH* mRNA will again adopt its secondary structure. By the combined action of sequestration of σ^{32} and reduced translation of its transcript, the heat shock response will be shut off even if the *E. coli* cells are kept at high temperatures.

5.1.2
Sigma-E

At elevated temperatures (43 °C and above), the transcription of *rpoH* becomes heat-inducible via the usage of a particular promoter, P3, and this P3 is not transcribed by the housekeeping σ^{70} factor. It has been shown that a new sigma factor, designated σ^E (or σ^{24}), is responsible for P3 recognition and transcription [55, 56]. The *rpoE* gene coding for σ^E was found to be the first gene of a four-gene operon designated *rpoE–rseA–resB–rseC* [57, 58]. The three distal genes play an important role in the signal transduction process coupled to the extracytoplasmic function of the σ^E-controlled regulon. This regulon is specifically induced in response to the imbalanced synthesis of outer membrane proteins [59] and by accumulation of misfolded proteins within the periplasm in general [60]. This induction is specific because it does not affect the cytoplasmic heat shock genes controlled by σ^{32}. Therefore, it can be concluded that the two heat shock regulons have evolved to respond to the accumulation of non-native proteins in the two different cellular compartments. The sigma-32 regulon senses non-native proteins in the cytoplasm, while the sigma-E regulon senses non-native proteins within the periplasm and the outer membrane.

What do we know about the regulation of the sigma-E regulon and of the function of the genes that are members of this regulon? A first clue came from the analysis of knockouts with the three *rse* genes (*rse* denotes *r*egulation of *s*igma *E*). While inactivation of *rseA* and *rseB* resulted in a high constitutive expression of σ^E, a *rseC* null mutation exhibited a weak down-regulatory effect pointing to RseA and RseB as negative and RseC as a positive regulator [57, 58]. A second clue came from the biochemical analysis of the Rse proteins. RseA was shown to directly interact with σ^E thereby classifying it as an anti-sigma factor. Furthermore, RseA was shown to be inserted in the cytoplasmic membrane in a fashion that positions the N-terminal domain in the cytoplasm, whereas the C-terminal domain protrudes into the periplasmic space. It is the N-terminal, cytoplasmic domain which binds σ^E [57, 58]. RseB interacts with the C-terminal, periplasmic domain of RseB and may function to modulate the RseA activity or to sense misfolded proteins [57, 61]. Finally, RseC, the fourth gene of this operon, is also an inner membrane protein, with unknown function, and could participate in regulating σ^E [58].

The current functional model of σ^E control is depicted in Fig. 2. In the absence of non-native proteins in the periplasm or outer membrane, σ^E is sequestered by RseA acting as an anti-sigma factor. Tight binding of σ^E requires the participation of RseB which might act as an co-anti-sigma factor. Upon accumulation of non-native proteins outside the cytoplasm, RseB is released from

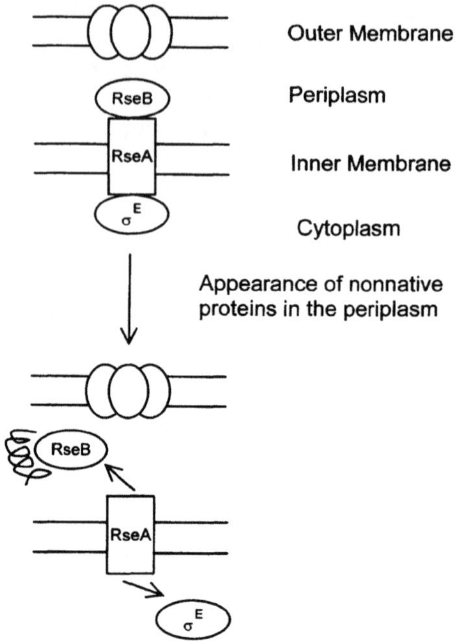

Fig. 2. Modulation of the RseA activity: a working model. In the absence of non-native proteins in the periplasm, σ^E is bound to the RseA anti-sigma factor; this binding is stabilized by RseB, a co-anti-sigma factor, in the periplasm. Upon accumulation of non-native proteins within the periplasm, RseB is released followed by dissociation of σ^E which interacts with RNA polymerase core enzyme to initiate transcription of the heat shock genes of the sigma-E regulon

RseA thereby generating a signal which is transduced to the N-terminal domain resulting in the dissociation of σ^E. The sigma factor will then bind to the RNA polymerase core enzyme, and the genes of the sigma-E regulon will become induced. What mechanism will ensure turn-off of the heat shock response? To answer this question, we can only speculate. Induction of the *sigE* operon will also result in an increase in the amount of RseA and RseB proteins. The exact function of RseB is not known but it is tempting to assume that it is related to the removal of non-native proteins. It might act as a chaperone, as a protease or exert some other function. When all the non-native proteins have been removed from the periplasm, RseB will interact with RseA which will then be able to sequester σ^E on the other side of the inner membrane.

As outlined above, many of the genes of the sigma-32 regulon are involved in the folding or degradation of heat-damaged proteins in the cytoplasm. One of the members of the sigma-E regulon, *degP* (also called *htrA*), codes for a periplasmic protein with two activities: general molecular chaperone and protease where a switch between these two activities is temperature-dependent, as mentioned above [42]. Another gene of this regulon, *fkpA*, encodes a peptidyl prolyl isomerase belonging to the FK506 binding protein family. This protein acts as a periplasmic-folding catalyst which can correct many periplasmic-folding defects [60]. All current knowledge on this topic has recently been reviewed [62].

5.1.3
Sigma-54

A third class of heat shock genes is under the control of the alternate sigma factor σ^{54} (for review, see [63]). It consists of only the five-gene operon *psp*, where *psp* denotes *p*hage-*s*hock-*p*rotein. The *psp* operon was discovered during the course of cloning of the filamentous phage IV gene since it induces the abundant synthesis of the PspA protein [64] and contains five open reading frames, at least four of which are expressed *in vivo* and *in vitro* [65]. This operon is expressed in response to a variety of environmental and intracellular stresses, among them heat shock. Since *psp* mutants are less able to maintain their membrane potential when cells are subjected to a specific, proton-motive force-depletion stress [66], it has been suggested that the *psp* genes constitute a stress-response operon which may respond to alterations in the energy charge of the cell. The observation that large amounts of PspA are produced upon strong stimulation led to the suggestion that this protein plays a major role and somehow prevents loss of protons from the cytoplasm.

Regulation of the *psp* operon is complex. As already mentioned, this operon is transcribed by σ^{54}, and this sigma factor needs an activator protein, encoded here by the *pspF* gene located upstream of the *psp* operon and transcribed in the opposite direction. The PspF activator binds to two sites located upstream of the σ^{54} promoter [67]. PspA has a dual function; besides preventing loss of protons from the cytoplasm, it also acts as a transcriptional repressor. PspB, C, and perhaps D, also have a role in regulating transcription of the *psp* operon. PspB and PspC act in concert as positive regulators, by relieving the repression imposed by PspA [65, 67, 68]. PspA resembles a coiled-coil protein, and it is likely that the inducing signal is transduced through protein–protein interactions by using these potentially interacting domains. It has to be stressed that induction of the *psp* operon by heat is independent of PspB, C and D [68]. In the absence of stress, PspA prevents expression of the *psp* operon, and it is an open question how repression of PspA is relieved. The simplest possibility would be that the protein itself is sensitive to temperature, as has been shown for the two DNA-binding proteins TlpA encoded by *Salmonella typhimurium* and the *Drosophila* heat shock factor HSF [69, 70].

5.1.4
Sigma-B

The alternative sigma factor σ^B, detected in 1979 as one of the first alternate sigma factors [71], is the master regulator of a large number of general stress genes induced in response to heat, ethanol, acid or salt stress as well as by starvation for glucose, phosphate and oxygen. The inducers of σ^B-dependent genes have been divided into two main groups [72, 73]. The first group contains extracellular signals that trigger a drop in the cellular ATP level. Starvation for oxygen, glucose or phosphate induces the *sigB* gene via ATP depletion. The second group of stress factors include heat, salt and ethanol. The cellular signal induced by these stimuli remains elusive, but does not involve a drop in the cellular ATP pool.

The *sigB* gene is encoded by the sigma-B operon which consists in total of eight genes in the order *rsbR–rsbS–rsbT–rsbU–rsbV–rsbW–sigB–rsbX*, where *rsb* denotes regulation of sigma *B*. A σ^A-dependent promoter in front of *rsbR* is responsible for the basal level of expression of all eight genes during vegetative growth [74], and a second internal promoter upstream of *rsbV* is recognized by σ^B itself and triggers expression of the last four genes of the *sigB* operon when induced by one of the stress factors mentioned. The activity of σ^B is controlled by the anti-sigma factor RsbW through direct protein–protein interactions. During vegetative growth in the absence of any stress factors both proteins form a stable complex which prevents σ^B interacting with RNA polymerase core enzyme. In response to stress or starvation, RsbW is attacked by another protein, RsbV, also designated as an anti-anti sigma factor. In the absence of stress, this RsbV is present in a phosphorylated form and therefore unable to attack RsbW. Stress and starvation enhance the level of non-phosphorylated RsbV, which is then able to bind to RsbW [75, 76]. The phosphorylation of RsbV is regulated by a complex signal transduction network composed of what has been designated as two "partner-switching modules". Each of the two modules contains three components, one RsbX-RsbS-RsbT and the other RsbU-RsbV-RsbW [77]. Both modules contain a serine phosphatase (RsbX, RsbU), an antagonist protein (RsbS, RsbV) and a switch protein/serine kinase (RsbT, RsbW). At present, an RsbU-independent and a RsbU-dependent pathway are distinguished. ATP depletion inhibits phosphorylation of RsbV by RsbW, the serine kinase [78, 79]. Therefore, the starvation pathway is RsbU-independent. Physical stress acts on the other module and stimulates the activity of the phosphatase RsbU which acts on RsbVP. To summarize, stress and starvation enhance the level of non-phosphorylated RsbV, which displaces the anti-sigma factor RsbW from σ^B which itself binds to the RNA polymerase core enzyme. Through an autoregulatory loop the level of σ^B increases resulting in a strong induction of the complete sigma-B regulon.

5.2
Regulation by Transcriptional Repressors

The fact that heat shock genes can also be subject to negative regulation by transcriptional repressors has become apparent quite recently. When our group started to study the heat shock response in *B. subtilis*, we first detected a novel regulatory sequence (termed CIRCE) which turned out to behave as a negative *cis*-acting element [80]. Then, a repressor gene and its operator were described as part of the *dnaK* operon of *Streptomyces coelicolor* [81]. Yet another putative repressor is used in *S. albus* to control expression of *hsp18* encoding a small HSP [82]. The potential repressor gene, *orfY*, is located 150 bp upstream of *hsp18* in the opposite orientation. Disruption of *orfY* results in elevated levels of *hsp18* transcript at low temperature, but this transcript is translated only after a heat shock indicating an additional post-transcriptional control. Lastly, a fourth, still unidentified, repressor must be present in *Bradyrhizobium japonicum* where at least five operons are under the control of the ROSE (*r*epression *o*f heat *s*hock gene *e*xpression) element. Very recently, a seond repressor encoding gene has been identified in *B. subtilis* and termed CtsR (*c*lass *t*hree *s*tress *r*epressor) [83]. Negative regulation of bacterial heat shock genes has recently been reviewed [84].

5.2.1
The HrcA Repressor

Analysis of the heat shock response in *B. subtilis* started with the cloning and sequencing of the *dnaK* and *groE* operons [85, 86]. Inspection of the upstream sequences revealed in both cases vegetative (σ^A-dependent) promoter sequences, and it was proven that in the case of the *groE* promoter it is indeed σ^A-dependent [87]. Visual inspection of the upstream regions further revealed two perfect inverted repeats each located between the transcriptional and translational start sites. To test whether these inverted repeats play a role during regulation of these two operons, one of them was mutated. Three bp each were exchanged in the left, the right and both arms, in the latter case restoring the dyad symmetry, of the inverted repeat located in front of the *dnaK* operon [80]. Subsequent analysis of the three mutant strains revealed that the genes of the *dnaK* operon were expressed at a high level comparable to what is observed after a heat shock. In contrast, expression of the *groE* operon was not affected by these mutations. The following conclusions were drawn: (i) the inverted repeat is indeed involved in the regulation of the downstream operons; (ii) the inverted repeat acts as a negative *cis*-element; (iii) dyad symmetry is not sufficient but rather the primary sequence of the inverted repeat suggesting that this sequence is recognized by a protein presumably a transcriptional repressor. Meanwhile, this inverted repeat has been described more than 80 times in more than 30 different bacterial species. In all these cases, it occurs upstream of either the *dnaK* and/or the *groE* operon, and since both operons encode molecular chaperons, it was designated as CIRCE (*c*ontrolling *i*nverted *r*epeat of *c*haperone *e*xpression).

Next, the gene coding for the repressor was identified by two different groups using two different experimental approaches. One group constructed a transcriptional fusion between the controlling region of the *groE* operon and a reporter gene coding for β-galactosidase. After mutagenesis of the whole *B. subtilis* chromosome, single colonies were screened for a blue phenotype at low temperature, and the mutations were mapped within the first gene of the *dnaK* operon [88]. The other group systematically inactivated all genes of the *dnaK* operon to find out whether they are essential or not. During this work, it turned out that deletion of the first gene resulted in a constitutive high level of expression of both the *dnaK* and the *groE* operon [89]. After its identification, the repressor gene was named *hrcA* (*h*eat *r*egulation at *C*IRCE).

How is the activity of the HrcA repressor modulated during a heat shock? The repressor must be modified in such a way to allow a transient high level of

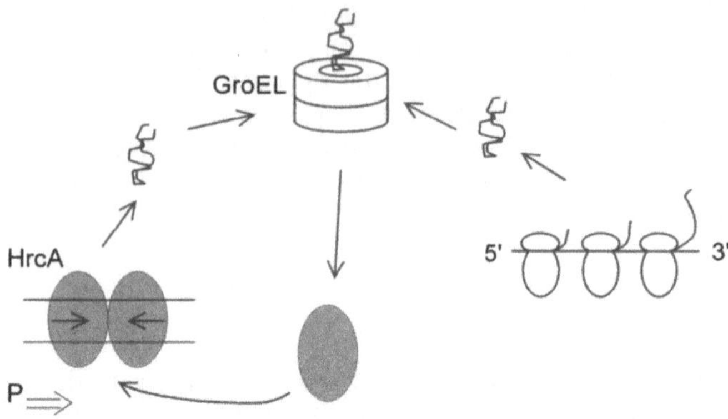

Fig. 3. Modulation of the activity of HrcA: a working model. It is assumed that HrcA repressor synthesized *de novo* or dissociated from its operator is present in an inactive form unable to interact with its operator. Through interaction with the GroE chaperone machine, inactive HrcA is converted into its active form

expression of the *dnaK* and *groE* operons. It was found that, in the absence of GroESL, the *dnaK* operon was constitutively expressed at a high rate in the presence of copious amounts of HrcA repressor [90]. In other words, the absence of GroE proteins resulted in the overproduction of inactive repressor protein. This and other observations led to the hypothesis that two different forms of HrcA repressor exist within the cell: inactive repressor, unable to bind to its operator, the CIRCE element, and active repressor, able to bind (Fig. 3). The conversion of the inactive to the active form needs the GroE chaperone system. It has been further suggested that *de novo* synthesized HrcA protein and protein molecules upon dissociation from their operator are present in the inactive form. This hypothesis nicely explains the transient behavior of the heat shock response. Upon a sudden temperature rise, the amount of non-native proteins will dramatically increase leading to a titration of the GroE machine. This in turn will result in the accumulation of inactive HrcA repressor. The more non-native proteins that have been removed from the cytoplasm the more GroE will be available to take care of HrcA thereby reactivating the repressor and turning off the heat shock response. It is still not known whether additional components are involved in modulating the activity of HrcA, e.g. non-native proteins might directly bind to HrcA to induce expression of *dnaK* and *groE* operons immediately after a heat shock.

5.2.2
The CtsR Repressor

The monocistronic genes *clpP* and *clpE* and the hexacistronic *clpC* operon of *B. subtilis* have been shown to be under the negative control of another repressor [83]. The product of the first gene of the *clpC* operon contains a helix-turn-helix motif and binds specifically to the promoter regions of all three *clp* genes. Since all these *clp* genes belong to the class III heat shock genes of *B. subtilis*, the gene encoding the repressor has been termed CtsR (*c*lass *t*hree *s*tress *r*epressor). The consensus CtsR recognition sequence (5′-A/GGTCAAA-3′) occurs as a directly repeated heptad upstream from the three *clp* genes. This CtsR box was also found upstream from *clp* and other heat shock genes of several Gram-positive species. In addition, *ctsR* orthologs were also identified in several of these bacteria. Concerning the mechanism of heat sensing by CtsR, it has been suggested that CtsR might act as a direct heat sensor, with intrinsic temperature sensitivity leading to reversible inactivation when exposed to elevated temperatures [83]. This hypothesis, however, does not explain the fact that the CtsR protein must regain activity even at high temperatures since expression of the *clp* genes is transient after a heat shock, as has been reported for all other heat shock genes. Since the activity of several heat shock regulators is modulated by molecular chaperones, and since it has been shown in *E. coli* that the Clp proteins ClpA and ClpX can act as chaperones, it is tempting to speculate that one of these two Clp proteins is involved in the regulation of the CtsR activity.

5.2.3
The HspR Repressor

While the *groESL*$_1$ operon and the *groEL*$_2$ gene of *Streptomyces coelicolor* and *S. albus* are controlled by HrcA/CIRCE, the *dnaK* operon is under the negative control of its own autoregulatory repressor [81]. This operon consists of the genes *dnaK–grpE–dnaJ–hspR* and is transcribed from a heat-inducible and growth phase regulated housekeeping promoter. The HspR repressor negatively controls transcription of the operon at 30 °C by interacting specifically with three inverted repeat motifs located upstream from the *S. coelicolor* and *S. albus dnaK* operons. Recently, the HspR binding site, now designated HAIR (*H*spR *as*sociated *i*nverted *r*epeat), has also been found upstream of the *clpB* gene [91]. A database search revealed that genes similar to *hspR* are also present in *Mycobacterium leprae*, *M. tuberculosis*, *Helicobacter pylori* and *Aquifex aeolicus* [91]. It is intriguing to speculate that the activity of the HspR repressor is modulated by the DnaK chaperon machine [90].

5.2.4
On the Search for a Repressor Interacting with the ROSE Element

In *Bradyrhizobium japonicum*, at least five heat shock operons are under the negative control of a still unidentified repressor. All five operons are preceded by an approx. 100-bp long regulatory sequence called ROSE (*r*epression *o*f heat *s*hock gene *e*xpression) which is located between the transcription and translation start sites of the first gene of each operon [92]. The ROSE element is mainly involved in controlling the regulation of heat shock genes that encode small heat shock proteins [93], but it also controls *rpoH*$_1$, one of the three genes coding for σ^{32} factors. The ROSE element does not seem to be widespread among bacterial species and does not occur in any of the totally sequenced bacterial genomes.

5.3 The Cellular Heat Shock Thermometer

The search for a cellular heat shock thermometer responsible for induction of the sigma-32 regulon dates back to 1990 when VanBogelen and Neidhardt discovered that both the heat and the cold shock could be induced by certain antibiotics [94]. They were clearly able to demonstrate that sub-lethal concentrations of the so-called H-group antibiotics (kanamycin, puromycin and streptomycin) induce a heat shock response. Since all these antibiotics leave the A site of the ribosomes empty, they suggested that the ribosomes act as the sensor (ribosome sensor model) and generate a signal which has not yet been identified. The biochemical nature of this signal could be (i) some small signal molecule analogous to ppGpp produced by the ribosomes in the H situation; (ii) the state of the ribosome itself might be the signal; (iii) the signal might be the nature of the proteins which are synthesized with decreased accuracy and thereby recognized by chaperones. A fourth possibility could be transient release of nascent polypeptide chains as an immediate response to heat shock.

A second suggestion was published a year later. Craig and Gross proposed that the free pool of DnaK serves as a cellular thermometer [95]. This suggestion is based on the observation that free σ^{32} is sequestered by DnaK/DnaJ under non-heat shock conditions and then either exposed to a protease such as FtsH or stored in association with DnaK thereby acting as an anti-sigma factor. Upon a sudden heat shock, the initial accumulation of misfolded proteins is thought to titrate free DnaK/DnaJ chaperones, resulting in stabilization of σ^{32}, whereas subsequent increase in HSPs destabilizes σ^{32} and represses its synthesis thereby forming an autogenous regulatory circuit. McCarty and Walker also suggested DnaK as a cellular thermometer [96]. They were able to show that DnaK autophosphorylates when incubated with ATP *in vitro*. The threonine-199 residue becomes phosphorylated which is critical for the ATPase activity of DnaK. Furthermore, the authors demonstrated that both the ATPase and autophosphorylating activities of DnaK increase very strongly between 20 and 53 °C. These observations led to the hypothesis that DnaK directly senses the environmental temperature, and this hypothesis does not demand that proteins unfold or misfold during temperature shifts.

Very recently, a third hypothesis has been published. Morita and co-workers [47] have suggested that the *rpoH* mRNA secondary structure itself acts as a thermosensor. In the absence of heat stress, the *rpoH* mRNA is folded into a secondary structure that occludes the ribosome binding site and the initiation codon. Upon heat shock, this structure is unfolded allowing ribosome binding and enhanced σ^{32} synthesis.

So we are left with four different potential cellular thermometers involved in inducing the sigma-32 regulon: the ribosomes, the DnaK/DnaJ chaperones, the ATPase activity of DnaK and the secondary structure of the *rpoH* mRNA. Any cellular thermometer has to fulfill the following requirements: (i) It must sense any increase in temperature even if it occurs, for example, from 30 to 34 °C; (ii) it must guarantee a rapid induction of the sigma-32 regulon; and (iii) it must further guarantee the transient nature of the heat shock response that means shut-off after a few minutes even if the cells are further incubated at the inducing temperature. The model of Craig and Gross [95] is attractive for explaining the induction of HSPs by various agents and treatments in the absence of any change in temperature which result in a slow steady state induction of heat shock gene expression. In contrast, a heat shock results in a rapid transient increase in heat shock gene expression which can be explained by the direct sensing of the environmental temperature by DnaK. The secondary structure model fails to explain shut-off of the sigma-32 regulon. Further experimentation is needed to identify the cellular thermometer, and it cannot be excluded that there is more than one thermometer depending on the stressor applied.

What do we know about the thermometers of the other heat shock regulons? There is only limited information. In the case of the sigma-E regulon, it has been suggested that RseB could sense non-native proteins within the periplasm [57] and, for the HrcA regulon, the GroE chaperone machine has been suggested to be involved in sensing the heat shock [90].

6 The Cold Shock Response

After a sudden downshift in the external temperature, bacterial cells have to deal with several problems: (1) secondary structures of RNA and DNA are stabilized which might affect the efficiencies of mRNA translation, DNA replication and recombination; (2) local melting of DNA by RNA polymerase presents a considerable problem for mesophilic bacteria thereby impairing transcription; and (3) the fluidity of the cytoplasmic membrane decreases which influences membrane-associated cellular functions such as active transport and protein secretion. To cope with the problems posed by cold stress, all bacterial species studied so far react with an adaptive response designated cold shock response. This response consists of sensing the temperature downshift, and generating a signal which will lead to the transient induction of a subset of chromosomal genes, the cold shock genes encoding CSPs. Immediately after cold shock, bacteria either completely or partly stop growth for a period of time, specific for the bacterial species. During this period, cells adapt to the lower temperature, and this lag-phase has been designated the acclimation phase. The cold shock response has been studied extensively in *Escherichia coli* (for recent reviews, see [97–99]) and to some extent in *Bacillus subtilis* (for recent reviews, see [100–102]). The influence of a temperature downshock on membrane fluidity has been analyzed in detail in cyanobacteria at the genetic level.

6.1 The Problem of Membrane Fluidity

The fluidity of the cellular membrane presents a limiting factor at low temperatures. After a decrease in temperature, membranes are too rigid and must therefore be desaturated. Normally, membranes are in a liquid crystalline form and will undergo a transition to a gel phase state when cells are exposed to a cold shock. Bacteria have developed mechanisms to compensate for the transition from the liquid crystalline to gel phase by changing the degree of saturation of the hydrocarbon chains of membrane phospholipids. Phospholipids with unsaturated fatty acids have lower melting points and a greater degree of flexibility than phospholipids containing saturated fatty acids. This type of response is known as homeoviscous adaptation [103, 104].

Fatty acid desaturases are the enzymes that are responsible for the introduction of specific double bonds into the hydrocarbon chains of fatty acids and, therefore, these enzymes play an important role during the process of cold acclimation. Two major mechanisms contribute to the alteration of the membrane phospholipid composition in response to temperature change: conversion of unsaturated to saturated fatty acids by preexisting but largely inactive desaturases or by *de novo* synthesis of these enzymes. In *E. coli*, the activity of a fatty acid synthesis protein converts palmitoleic to *cis*-vaccenic acid within 30 s following a cold shock [105]. The enzyme is present at all temperatures but is more active at lower temperatures [106]. In *Bacilli*, a temperature downshift results in increased synthesis of membrane-bound desaturase and enhanced stability of this enzyme [107]. When *B. subtilis* is grown at 37 °C, cells synthesize, almost exclusively, saturated fatty acids. When the cells were transferred from 37 to 20 °C, synthesis of a C-16 unsaturated fatty acid was observed, which reached a plateau after 5 h at 20 °C, reaching levels of 20% of the total fatty acid content [108]. Recently, the gene encoding this enzyme has been identified [109]. It has been designated *des* and codes for a Δ5 acyl-lipid desaturase. Upon disruption of the *des* gene, cells were unable to synthesize unsaturated fatty acids, exhibited growth characteristics comparable to its congenic parent strain at 37 and 20 °C, but showed severely reduced survival during the stationary phase.

Cyanobacteria, prokaryotic algae that perform oxygenic photosynthesis, respond to a decrease in ambient growth temperature by desaturating the fatty acids of membrane lipids to compensate for the decrease in the molecular motion of the membrane lipids at low temperatures. During low-temperature acclimation of cyanobacterial cells, the desaturation of fatty acids occurs without *de novo* synthesis of fatty acids [110, 111]. All known cyanobacterial desaturases are intrinsic membrane proteins that act on acyl-lipid substrates.

A shift in temperature from 38 to 22 °C leads to desaturation of fatty acids in *Anabaena variabilis* [110], resulting in control of the fluidity of the plasma membrane. Mutants have been isolated in *Synechocystis* PCC 6803 that were defective in desaturation of fatty acids, and the growth rate of one of these mutants was much lower than that of the wild-type at 22 °C [112]. It turned out that the mutant strain had a mutation in the gene *desA*, and when the wild-type allele was introduced into the chilling-sensitive cyanobacterium *Anacystis nidulans*, it resulted in increasing the tolerance of that strain to low temperature [113]. These experiments nicely demonstrate the existence of a mechanism of adaptation to low temperature in a chilling-tolerant cyanobacterium.

In *Synechococcus* sp. strain PCC 7002, the temperature-regulated mRNA accumulation of the three desaturase genes, *desA* (Δ12 desaturase), *desB* (ω3 desaturase) and *desC* (Δ9 desaturase), have been studied in detail and have revealed three patterns of temperature regulation [114].

6.2
The Cold Shock Response in *E. coli*

When *E. coli* cells are transferred from 37 to 10 °C, cells stop growth immediately and protein synthesis is inhibited for about 4 h followed by resumed growth with a generation time of 24 h [115, 116]. During this acclimation phase many physiological changes occur, including a decrease in the saturation of fatty acids and inhibition of DNA, RNA and protein synthesis. Inhibition of growth seems to be primarily the consequence of translation inhibition. Here, the formation of temperature-dependent secondary structures near the 5′ end of mRNA molecules which lower the access of ribosomes to their Shine-Dalgarno sequence is discussed as the major reason for reduced translation. During the acclimation phase, only a specific class of proteins is synthesized, the CSPs, that can be detected by two-dimensional gel electrophoresis [116–119], and these CSPs play an important role in the adaptation of the cell to growth at low temperatures especially in translation initiation (see below). At the end of the acclimation phase, the synthesis of these CSPs becomes repressed, and cells resume growth.

In contrast to *E. coli*, *B. subtilis* cells do not enter a 4-h lag phase. Instead, they continue growth with a generation time of more than 24 h and induce the synthesis of 36 CSPs which function at various levels of cellular physiology, such as protein synthesis, carbohydrate uptake and chemotaxis [120].

6.2.1
Major E. coli Cold Shock Proteins and Their Function

In *E. coli*, two classes of cold shock genes are distinguished by their induction ratio. While class I genes are induced at least 10-fold, those of class II are induced only a few-fold. Genes which are induced 10-fold and more include four members of the Csp family (see next chapter) and the three genes *csdA*, *rbfA* and *hscA*. Gene *csdA* codes for a 70-kDa protein which can be detected associated with ribosomes in cells grown at 15 °C. Purified CsdA was found to unwind double-stranded RNA in the absence of ATP and is also called DeaD box RNA helicase [119]. In the absence of CsdA, growth of the cells and synthesis of a number of proteins, among them major heat shock proteins including DnaK and GroEL, was impaired. All these results led to the hypothesis that CsdA plays an essential role in mRNA translation at low temperatures by unwinding stable secondary structures in mRNA and thereby allowing ribosomes to bind to their Shine-Dalgarno sequence. Another major CSP is the 15-kDa protein RbfA, which plays a role in ribosomal maturation and/or initiation of translation [121]. The general role attributed to RbfA at low temperatures is to overcome the inhibition of translation. The role of the HscA is less clear. Since it exhibits homology to the DnaK chaperone, it might act in a similar way as a molecular chaperone during cold shock.

6.2.2
The Csp family of Related Cold Shock Proteins

Most pronounced within the cold shock response is the high level of induction of a CSP designated CspA (~7.5 kDa) [122]. This protein is a member of a family of Csp's widespread in many bacterial species. In *E. coli*, this family has nine members, and four *csp* genes are cold-inducible (*cspA*, *cspB*, *cspG* and *cspI*). The Csp's share more than 43% identity with the domain of the eukaryotic Y-box proteins called the cold shock domain (CSD) which plays an important role in gene regulation and mRNA masking by binding to the Y-box sequence ATTGG [123]. The CspA protein is not only a cold shock protein, but is also synthesized in massive amounts during early exponential growth at 37 °C and therefore plays an important role during growth cycle regulation [124]. The *cspA* mRNA level decreases rapidly with increasing cell density, becoming virtually undetectable by mid-to-late exponential growth phase. Therefore, the extent of CspA induction depends on its preshock level.

Most of the work has concentrated on the structure and function of CspA. Its structure has been determined both by NMR and X-ray crystallography [125, 126]. CspA contains five anti-parallel β-sheets, which form a β-barrel structure. Two RNA-binding motifs are located on a β-2- and a β-3-sheet, respectively, suggesting potential interactions between CspA and nucleic acids (RNA and DNA). Three different functions have been reported for CspA. First, it was proposed to function as an RNA chaperone that facilitates translation of cellular mRNAs by destabilizing mRNA secondary structures formed at low temperatures [127]. Second, CspA acts as a transcriptional activator protein, and, third, CspA is involved in shutting down its own synthesis at 37 °C by acting as a negative regulator protein (see below).

In *B. subtilis*, three *csp* genes have been identified which are all induced by a temperature downshock; they are desiganted *cspB*, *cspC* and *cspD*. The products of these genes are required under different physiological conditions. While disruption of *cspB* caused a freezing-sensitive phenotype and has been shown to affect the level of induction of several cold-inducible and other proteins after cold shock [120], deletion of *cspC* and *cspD* genes did not result in a detectable phenotype [101]. In contrast, *csp* double mutants exhibited severe reduction in cellular growth at 15 and 37 °C, including impairment of survival during the stationary phase. Furthermore, protein synthesis was deregulated in *csp* double mutants and the loss of one or two Csp proteins led to an increase in the synthesis of the remaining *csp* gene(s) [101]. Interestingly, the presence of at least one *csp* is essential for viability in *B. subtilis*, not only at low temperatures, but also at 37 °C [101].

The *csp* genes were found in the earliest diverging bacterial branches of *Aquifex* and *Thermotoga*, but not in the archael genomes that have been entirely sequenced so far. In addition, the chromosomes of the eubacteria *Synechocystis* sp. PCC6803 and *Borrelia burgdorferi* do not contain *csp*-like sequences. Interestingly, cyanobacteria possess a family of small cold-inducible proteins (Rbps) that are homologous to the RNA-binding domain that is present in eukaryotic RNA-binding proteins [128]. As an *rbd*-like sequence is not found in

Gram-positive or other -negative bacteria, *rbp* genes could have been replaced by *csp* genes in cyanobacteria.

6.2.3
Other Cold Shock Proteins

Among the CSPs which are only moderately induced are NusA (involved in termination and antitermination), RecA (DNA recombination and SOS response), H-NS (a global regulator with many binding sites on the chromosome), GyrA (the α-subunit of the topoisomerase gyrase), initiation factor 2 of translation (subunits a and b), polynucleotide phosphorylase (degradation of mRNA), pyruvate dehydrogenase, dihydrolipoamide acetyltransferase (a subunit of pyruvate dehydrogenase) and the trigger factor (molecular chaperone with prolylisomerase activity). Most of these proteins interact with DNA or RNA, and a role in resolving constraints imposed by the temperature downshock can be envisaged. There are no publications dealing specifically with these minor cold shock proteins and their function during the acclimation phase.

6.3
Regulation of the Cold Shock Response

In contrast to the heat shock response, there is only limited information on the regulation of the cold shock response, and this exists only for *E. coli*. Most publications deal with the regulatory role of CspA after a temperature downshock.

What could be the signal for the induction of the cold shock proteins? It has been observed that shifting *E. coli* cells from 37 to 5 °C results in an accumulation of 70S monosomes with a concomitant decrease in the number of polysomes [129]. Further, it has been shown that a cold shock response is induced when ribosomal function is inhibited, e.g. by cold-sensitive ribosomal mutations [121] or by certain antibiotics such as chloramphenicol [94]. These data indicate that the physiological signal for the induction of the cold shock response is inhibition of translation caused by the abrupt shift to lower temperature. Then, the cold shock proteins RbfA, CsdA and IF2 associate with the 70S ribosomes to convert the cold-sensitive nontranslatable ribosomes into cold-resistant translatable ribosomes. This in turn results in an increase in cellular protein synthesis and growth of the cells.

In contrast to most mRNAs, which become untranslatable after a temperature downshock, cold shock mRNAs possess a mechanism to form the translation initiation complex at low temperature without cold shock ribosomes. A close inspection of the mRNAs of class I cold shock proteins reveal that they are equipped with an extra ribosome-binding site called the downstream box located within the coding region of their transcript [130]. It would be interesting to know whether introduction of this downstream box into a cellular mRNA would convert it into a transcript which can be transcribed immediately after a cold shock. In the case of the *cspA* mRNA it has been shown that in the absence of the downstream box the initiation complex cannot be formed at low temperature during the acclimation phase [131].

6.3.1
Regulation of cspA

As already mentioned, there are in total nine *csp* genes encoded by the *E. coli* chromosome, where four are induced by a cold shock (*cspA*, *cspB*, *cspG* and *cspI*) and the maximal induction of each gene is dependent on the growth temperature. The almost identical amino acid residue composition of the four Csp proteins makes it difficult to distinguish between them, and the only experimental technique which fulfills this requirement uses the separation of the Csp proteins by 2D gel electrophoresis. Since CspA was the first major cold shock protein to be described and CspI has only been detected very recently, all the work has concentrated on CspA. It should be mentioned again that the *cspA* gene is not only induced by heat as thought, but is also produced in large amounts during the early exponential phase [124]. However, we will consider here only what is known about its regulation after a temperature downshock.

Regulation of *cspA* occurs at three different stages: in the absence of cold shock, immediately after a temperature downshift and at the end of the acclimation phase. In *E. coli* cells growing experimentally at 37 °C, the *cspA* promoter is highly active most probably because of the UP element, an AT-rich sequence immediately upstream of the –35 region, but no CspA protein is present [130, 132, 133]. The core *cspA* promoter (from – 40 to + 16) responds to cold shock and a mutation at –36 increases the relative activity of the promoter at low temperatures three-fold [132], and an unidentified factor has been found to bind upstream of the core promoter [134].This observation suggests that expression of *cspA* is regulated post-transcriptionally. One facet of this post-transcriptional regulation involves stability of the *cspA* mRNA. It has been shown that the transcript is extremely unstable at 37 °C but is dramatically stabilized upon cold shock [130, 134–136]. The *cspA* mRNA (and that of other major cold shock genes) possesses an unusually long 5′-untranslated region (designated 5′-UTR) consisting of 159 bases [134], and this 5′-UTR plays a crucial role in its cold inducibility [130, 133]. Immediately downstream of all four cold-inducible *csp* genes, a nearly identical sequence is present which has been designated the cold box sequence (Table 4). Recently, it has been shown that the product of the *cspE*

Table 4. Cold box present near the 5′ end of major *E. coli* cold shock genes

	–10 region[a]	+1[b]	Cold box
cspA	CTTAAT	GCACATCAA-C	**GGTTTGA**CGTACAGAC
cspB	GTTAAT	-GATTGC-GTC	**GGTTTGAAGAACAGAC**
cspG	GTTAAT	-AGCTGC-GTC	**GGTTTGAAAGACAGAC**
cspI	GTTAAT	GGTGTT-CT-	**GGTTTGT**TACAAATTT
hscA		GG-	**GGTTTG**TTTTACCTGA
Consensus		**GGTTTGA**	

[a] The – 10 region indicates the second half of the potential promoter sequence recognized by the vegetative sigma factor.

[b] + 1 indicates the mapped potential putative transcription start site. The cold box is shown in boldface.

gene might interact with the transcription elongation complex in the *cspA* cold box region [135]. In addition, translation of the *cspA* mRNA seems to be very efficient at low temperatures in comparison with mRNAs of non-cold-shock proteins [130].

After a cold shock, the magnitude of enhancement at the transcriptional level is quantitatively modest, and it is considered that the cold induction is mainly controlled at steps after transcription initiation. Stabilization of the *cspA* transcript is a major cause for the cold induction of CspA because the *cspA* mRNA is extremely unstable *in vivo* at 37 °C, with a half-life of about 10 s, and becomes dramatically stabilized immediately after a cold shock [136, 137]. Three base substitutions near the Shine-Dalgarno sequence caused a 150-fold enhancement of *cspA* transcript stability, resulting in constitutive expression of *cspA* at 37 °C [133]. Both *cspA* mRNA destabilization at 37 °C and stabilization after cold shock were recently shown to depend on the 5′-UTR [130]. Control at the level of translation is exerted by the so-called downstream box in the *cspA* coding sequence [130]. This downstream box, that is complementary to a sequence proximal to the ribosome-binding site decoding region in 16S rRNA, is required for efficient translation under cold shock conditions [130]. A downstream box was also found in other *E. coli* cold shock genes and could therefore represent a general mechanism for translational cold shock induction.

Repression of *cspA* after acclimation at low temperatures appears to be mediated by the cold box region and is a possible target site for repressor binding [138]. In addition, CspA itself negatively regulates its own gene expression through the cold box [138, 139].

6.3.2
Regulation of the Other Cold Shock Genes

As already mentioned, CspA acts as a transcriptional activator by binding to the CCAAT sequence starting at +12 of a 110-bp fragment including the promoter of the *hns* gene coding for a global repressor. Upon cold shock, there is an approx. 4-fold increase in the cellular levels of both *hns* mRNA and H-NS protein [140]. Overproduction of CspA during cold shock facilitates the synthesis of at least three additional cold shock proteins, among them GyrA [117]. Here, the *gyrA* promoter contains three ATTGG sequences which are spaced two and three integral turns of DNA apart, and CspA forms three complexes with the promoter region. Taking these data together, CspA might have a general role in enhancing transcription under cold shock conditions through binding at the common ATTGG or its palindrom CCAAT element found in several diverse promoters.

6.4
The Cellular Cold Shock Thermometer

It has been suggested that the ribosome might act as the sensor for the cold shock response [94, 99], and it was postulated that the physiological signal for the induction of the cold shock response may be inhibition of initiation of translation caused by the temperature downshift. Three ribosome-associated proteins, IF2, CsdA and RbfA, are induced during the transient block in initiation of translation in *E. coli* as outlined above to convert the cold-sensitive non-translatable ribosome into a cold-resistant translatable state. At least one of these proteins, CsdA, is also involved in optimal growth at low temperatures, possibly by unwinding stable secondary structures in mRNA [119].

Acknowledgement. The author would like to thank the DFG and the Fonds der Chemischen Industrie for continued financial support.

References

1. Harder W, Veldkamp H (1967) Arch Mikrobiol 59:123
2. Herendeen SL, VanBogelen RA, Neidhard FC (1979) J Bacteriol 139:185
3. Condon C, Liveris D, Squires C, Schwartz I, Squires CL (1995) J Bacteriol 177:4152
4. Bukau B, Horwich AL (1998) Cell 92:351
5. Bohen SP, Kralli A, Yamamoto KR (1995) Science 268:1303
6. Rassow J, Von Ahsen O, Bömer U, Pfanner N (1997) Trends Cell Biol 7:129
7. Richardson A, Landry SJ, Georgopoulos C (1998) Trends Biochem Sci 23:138
8. Wickner S, Hoskins J, McKenney K (1991) Proc Natl Acad Sci USA 88:7903
9. Wickner S, Gottesman S, Skowyra D, Hoskins J, McKenney K, Maurizi MR (1994) Proc Natl Acad Sci USA 91:12218
10. Wawrzynow A, Wojtkowiak D, Marszalek J, Banecki B, Jonsen M, Graves B, Georgopoulos C, Zylicz M (1995) EMBO J 14:1867
11. Konieczny I, Helinski DR (1997) Proc Natl Acad Sci USA 94:14378
12. Levchenko I, Luo L, Baker TA (1995) Genes Dev 9:2399
13. Kruklitis R, Welty DJ, Nakai H (1996) EMBO J 15:935
14. Laskowska E, Kuczynska-Wisnik D, Skorko-Glonek J, Taylor A (1996) Mol Microbiol 22:555
15. Jakob U, Muse W, Eser M, Bardwell JCA (1999) Cell 96:341
16. Allen SP, Polazzi JO, Gierse JK, Easton AM (1992) J Bacteriol 174:6938
17. Laskowska E, Wawrzynów A, Taylor A (1996) Biochimie 78:117
18. Jakob U, Buchner J (1994) Trends Biochem Sci 19:205
19. Bardwell JCA, Craig EA (1987) Proc Natl Acad Sci USA 84:5177
20. Jakob U, Lilie H, Meyer I, Buchner J (1995) J Biol Chem 270:7288
21. Jakob U, Meyer I, Bügl H, André S, Bardwell JCA, Buchner J (1995) J Biol Chem 270: 14412
22. Versteeg S, Mogk A, Schumann W (1999) Mol Gen Genet 261:582
23. Goldberg AL, Moerschell RP, Chung CH, Maurizi MR (1994) Methods Enzymol 244:350
24. Gottesman S (1996) Annu Rev Genet 30:465
25. Gottesman S (1996) Roles for energy-dependent proteases in regulatory cascades. In: Lin ECC, Lynch AS (eds) Regulation of gene expression in *Escherichia coli*. R.G. Landes Co., Austin, TX, p 503
26. Maurizi MR, Trisler P, Gottesman S (1985) J Bacteriol 164:1124
27. Kessel M, Maurizi MR, Kim B, Kocsis E, Trus BL, Singh SK, Steven AC (1995) J Mol Biol 250:587
28. Gottesman S, Clark WP, De Crecy-Lagard V, Maurizi MR (1993) J Biol Chem 268:22618
29. Wojtkowiak D, Georgopoulos C, Zylicz M (1993) J Biol Chem 268:22609
30. Van Melderen L, Thi MHD, Lecchi P, Gottesman S, Couturier M, Maurizi MR (1996) J Biol Chem 271:27730
31. Missiakas D, Schwager F, Betton J-M, Georgopoulos C, Raina S (1996) EMBO J 15:6899
32. Kessel M, Wu WF, Gottesman S, Kocsis E, Steven AC, Maurizi MR (1996) FEBS Lett 398:274

33. Rohrwild M, Coux O, Huang HC, Moerschell RP, Yoo SJ, Seol JH, Chung CH, Goldberg AL (1996) Proc Natl Acad Sci USA 93:5808
34. Schumann W (1999) FEMS Microbiol Rev 23:1
35. Tomoyasu T, Yuki T, Morimura S, Mori H, Yamanaka K, Niki H, Hiraga S, Ogura T (1993) J Bacteriol 175:1344
36. Tomoyasu T, Gamer J, Bukau B, Kanemori M, Mori H, Rutman AJ, Oppenheim AB, Yura T, Yamanaka K, Niki H, Hiraga S, Ogura T (1995) EMBO J 14:2551
37. Kihara A, Akiyama Y, Ito K (1995) Proc Natl Acad Sci USA 92:4532
38. Akiyama Y, Kihara A, Ito K (1996) FEBS Lett 399:26
39. Kihara A, Akiyama Y, Ito K (1998) J Mol Biol 279:175
40. Akiyama Y, Ehrmann M, Kihara A, Ito K (1998) Mol Microbiol 28:803
41. Rawlings N, Barrett A (1994) Methods Enzymol 244:19
42. Spiess C, Beil A, Ehrmann M (1999) Cell 97:339
43. Grossman AD, Erickson JW, Gross CA (1984) Cell 38:383
44. Bukau B (1993) Mol Microbiol 9:671
45. Yura T, Nagai H, Mori H (1993) Annu Rev Microbiol 47:321
46. Nagai H, Yuzawa H, Yura T (1991) Proc Natl Acad Sci USA 88:10515
47. Morita MT, Tanaka Y, Kodama TS, Kyogoku Y, Yanagi H, Yura T (1999) Genes Dev 13:655
48. Nagai H, Yuzawa H, Kanemori M, Yura, T (1994) Proc Natl Acad Sci USA 91:10280
49. Nakahigashi K, Yanagi H, Yura T (1995) Nucleic Acids Res 23:4383
50. McCarty JS, Rüdiger S, Schönfel HJ, Schneider-Mergener J, Nakahigashi K, Yura T, Bukau B (1996) J Mol Biol 256:829
51. Straus DB, Walter WA, Gross CA (1987) Nature 329:348
52. Tilly K, Spence J, Georgopoulos C (1989) J Bacteriol 171:1585
53. Straus DB, Walter W, Gross CA (1990) Genes Dev 4:2202
54. Herman C, Thévenet D, D'Ari R, Bouloc P (1995) Proc Natl Acad Sci USA 92:3516
55. Erickson JW, Gross CA (1989) Genes Dev 3:1462
56. Wang QP, Kaguni JM (1989) J Bacteriol 171:4248
56. De Las Peñas A, Connolly L, Gross CA (1997) Mol Microbiol 24:373
58. Missiakas D, Mayer MP, Lemaire M, Georgopoulos C, Raina S (1997) Mol Microbiol 24:355
59. Mecsas J, Rouviere PE, Erickson JW, Donohue TJ, Gross CA (1993) Genes Dev 7:2618
60. Missiakas D, Betton JM, Raina S (1996) Mol Microbiol 21:871
61. Raina S, Missiakas D, Georgopoulos C (1995) EMBO J 14:1043
62. Missiakas D, Raina S (1998) Mol Microbiol 28:1059
63. Model P, Jovanovic G, Dworkin J (1997) Mol Microbiol 24:255
64. Brissette JL, Russel M, Weiner L, Model P (1990) Proc Natl Acad Sci USA 87:862
65. Brissette JL, Weiner L, Ripmaster TL, Model P (1991) J Mol Biol 220:35
66. Kleerebezem M, Crielaard W, Tommassen J (1996) EMBO J 15:162
67. Weiner L, Brissette JL, Ramani N, Model P (1995) Nucleic Acids Res 23:2030
68. Weiner L, Brissette JL, Model P (1991) Genes Dev 5:1912
69. Hurme R, Berndt KD, Normark SJ, Rhen M (1997) Cell 90:55
70. Zhong M, Orosz A, Wu C (1998) Mol Cell 2:101
71. Haldenwang WG, Losick R (1979) Nature 282:256
72. Alper S, Duncan L, Losick R (1994) Cell 77:195
73. Kang CM, Brody MS, Akbar S, Yang XF, Price CW (1996) J Bacteriol 178:3846
74. Wise AA, Price CW (1995) J Bacteriol 177:123
75. Dufour A, Haldenwang WG (1994) J Bacteriol 176:1813
76. Voelker U, Voelker A, Haldenwang WG (1996) J Bacteriol 178:5456
77. Yang X, Kang CM, Brody MS, Price CW (1996) Genes Dev 10:2265
78. Voelker U, Voelker A, Maul B, Hecker M, Dufour A, Haldenwang WG (1995) J Bacteriol 177:3771
79. Alper S, Dufour A, Garsin DA, Duncan L, Losick R (1996) J Mol Biol 260:165
80. Zuber U, Schumann W (1994) J Bacteriol 176:1359
81. Bucca G, Ferina G, Puglia AM, Smith CP (1995) Mol Microbiol 17:663

82. Servant P, Mazodier P (1995) J Bacteriol 177:2998
83. Derré I, Rapoport G, Msadek T (1999) Mol Microbiol 31:117
84. Narberhaus F (1999) Mol Microbiol 31:1
85. Wetzstein M, Völker U, Dedio J, Löbau S, Zuber U, Schiesswohl M, Herget C, Hecker M, Schumann W (1992) J Bacteriol 174:3300
86. Schmidt A, Schiesswohl M, Völker U, Hecker M, Schumann W (1992) J Bacteriol 174:3993
87. Yuan G, Wong S-L (1995) J Bacteriol 177:5427
88. Yuan G, Wong S-L (1995) J Bacteriol 177:6462
89. Schulz A, Schumann W (1996) J Bacteriol 178:1088
90. Mogk A, Homuth G, Scholz C, Kim L, Schmid FX, Schumann W (1997) EMBO J 16:4579
91. Grandvalet C, de Crécy-Lagard V, Mazodier P (1999) Mol Microbiol 31:521
92. Narberhaus F, Käser R, Nocker A, Hennecke H (1998) Mol Microbiol 28:315
93. Münchbach M, Nocker A, Narberhaus F (1999) J Bacteriol 181:83
94. VanBogelen RA, Neidhardt FC (1990) Proc Natl Acad Sci USA 87:5589
95. Craig EA, Gross CA (1991) Trends Biochem Sci 16:135
96. McCarty JS, Walker GC (1991) Proc Natl Acad Sci USA 88:9513
97. Jones PG, Inouye M (1994) Mol Microbiol 11:811
98. Yamanakada K, Fang L, Inouye M (1998) Mol Microbiol 27:247
99. Thieringer HA, Jones PG, Inouye M (1998) BioEssays 20:49
100. Graumann P, Marahiel MA (1996) Arch Microbiol 166:293
101. Graumann P, Wendrich TM, Weber MHW, Schröder K, Marahiel MA (1997) Mol Microbiol 25:741
102. Graumann PL, Marahiel MA (1998) Trends Biochem Sci 23:286
103. Murata N (1989) J Bioenerg Biomembr 21:61
104. Cossins AR (1994) In: Cossins AR (ed) Temperature adaptation of biological membranes. Portland, London, p 63
105. Garwin JL, Cronan JE Jr (1980) J Bacteriol 141:1457
106. Cronan JE Jr, Rock CO (1996) In: Neidhard FC, Curtiss R III, Ingraham JL, Lin ECC, Low KB, Magasanik B, Reznikoff WS, Riley M, Schaechter M (eds) *Escherichia coli* and *Salmonella*: cellular and molecular biology, 2nd ed. ASM Press, Washington DC, p 612
107. Fulco AJ, Fujii DK (1980) The effect of temperature on the formation of delta-5 unsaturated fatty acids by bacilli. In: Kates M, Kuksis A (eds) Membrane fluidity, biophysical techniques and cellular regulation. Humana Press, Totowa, New Jersey, p 77
108. Grau R, De Mendoza D (1993) Mol Microbiol 8:535
109. Aguilar PS, Cronan JE Jr, De Mendoza D (1998) J Bacteriol 180:2194
110. Sato N, Murata N (1981) Plant Cell Physiol 22:1043
111. Wada H, Murata N (1990) Plant Physiol 92:1062
112. Wada H, Murata N (1989) Plant Cell Physiol 30:971
113. Wada H, Gombos Z, Murata N (1990) Nature 347:200
114. Sakamoto T, Bryant DA (1997) Mol Microbiol 23:1281
115. Ng H, Ingraham JL, Marr AG (1962) J Bacteriol 84:331
116. Jones PG, VanBogelen RA, Neidhardt FC (1987) J Bacteriol 169:2092
117. Jones PG, Krah R, Tafuri SR, Wolffe AP (1992) J Bacteriol 174:5798
118. Lee SJ, Xie A, Jiang W, Etchegary J-P, Jones PG, Inouye M (1994) Mol Microbiol 11:833
119. Jones PG, Mitta M, Kim Y, Jiang WN, Inouye M (1996) Proc Natl Acad Sci USA 93:76
120. Graumann P, Schröder K, Schmid R, Marahiel MA (1996) J Bacteriol 178:4611
121. Jones PG, Inouye M (1996) Mol Microbiol 21:1207
122. Goldstein J, Pollitt NS, Inouye M (1990) Proc Natl Acad Sci USA 87:283
123. Wolffe AP (1994) BioEssays 16:245
124. Brandi A, Spurio R, Gualerzi CO, Pon CL (1999) EMBO J 18:1653
125. Newkirk K, Feng W, Jiang W, Tejero R, Emerson SD, Inouye M, Montelione GT (1994) Proc Natl Acad Sci USA 91:5114
126. Schindelin H, Jiang W, Inouye M, Heinemann U (1994) Proc Natl Acad Sci USA 91:5119
127. Jiang W, Hou Y, Inouye M (1997) J Biol Chem 272:196

128. Sato N (1995) Nucleic Acids Res 23:2161
129. Broeze RJ, Solomon CJ, Popham D (1978) J Bacteriol 134:861
130. Mitta M, Fang L, Inouye M (1997) Mol Microbiol 26:321
131. Jiang W, Jones P, Inouye M (1993) J Bacteriol 175:5824
132. Goldenberg D, Azar I, Oppenheim AB, Brandi A, Pon CL, Gualerzi CO (1997) Mol Gen Genet 256:282
133. Tanabe H, Goldstein J, Yang M, Inouye M (1992) J Bacteriol 174:3867
134. Fang L, Jiang W, Bae W, Inouye M (1997) Mol Microbiol 23:355
135. Brandi A, Pietroni P, Gualerzi CO, Pon CL (1997) Mol Microbiol 19:231
136. Goldenberg D, Azar I, Oppenheim AB (1996) Mol Microbiol 19:241
137. Bae W, Phadtare S, Severinov K, Inouye M (1999) Mol Microbiol 31:1429
138. Jiang W, Fang L, Inouye M (1996) J Bacteriol 178:4919
139. Bae W, Jones PG, Inouye M (1997) J Bacteriol 179:7081
140. La Teana A, Brandi A, Falconi M, Spurio R, Pon CL, Gualerzi CO (1991) Proc Natl Acad Sci USA 88:10907

Received August 1999

Particle Stress in Bioreactors

Hans-Jürgen Henzler

Bayer AG, Process Development/Bioprocess Engineering, D-42096 Wuppertal, Germany
E-mail: hans-juergen.henzler.hh@bayer-ag.de

In many biological processes, e.g. the fermentation of cells and sensitive microorganisms or bioconversion with immobilised enzymes, low shear stress is of crucial importance for the optimal course of processes.

Starting with the causes of particle stress, the following report discussed the hydrodynamic principles of the most frequently used model reactors and bioreactors, which are required for an approximate calculation of stress.

The main part of the report describes the results of systematic investigations into the hydrodynamic stress on particles in stirred tanks, reactors with dominating boundary-layer flow, shake flasks, viscosimeters, bubble columns and gas-operated loop reactors. These results for model and biological particle systems permit fundamental conclusions on particle stress and the dimensions and selection of suitable bioreactors according to the criterion of particle stress.

Advances in Biochemical Engineering/
Biotechnology, Vol. 67
Managing Editor: Th. Scheper

Abbreviations

a	Enzyme activity
c	Concentration
c_F	q_F/nd^3, Pumping number of impeller
d	Impeller diameter
d_E	Droplet diameter
d_F	Mean floc diameter
d_{Fv}	Reference floc diameter: Diameter at $d(d_F)/dt = 0.0055$ [mm/s]
d_L	Hole diameter of gas distributor
d_P	Particle diameter
d_S	Maximum inside flask diameter
D	Tank diameter or tube diameter or height of rectangular channel
Fr	n^2d/g, Froude number
g	Acceleration due to gravity
h	Vertical height of impeller blade
h_B	Distance impeller to bottom of reactor
H	Fill height

k_d	$1/t \cdot \ln(N_o/N)$, death rate
k_La	Gas-liquid mass transfer coefficient
K	Consistency index
m	Flow exponent
n	Revolutions of impeller or shaker machine
N	Number of cells at actual time
N_o	Number of cells at the beginning of stress experiments
Ne	$P/\varrho n^3 d^5$, Newton number for impeller systems
Ne_S	$P/\varrho n^3 d_S^5$, Newton number for shake flasks
Ne*	Newton number for viscosimeter, corresponding to Eq. (12)
p	Pressure
p_o	Environmental pressure
P	Power input
P_a	Adiabatic power
P/Xt	Productivity, relation of product P to cell mass concentration X and time t
q	Gas throughput
q_F	Pumping throughput of impeller
Q	q/nd^3, gas throughput number
r	Radial coordinate
r_1	Diameter of the inner cylinder of viscometer
r_2	Diameter of the outer cylinder of viscometer
Re	Reynolds number
t	Time
Ta	Taylor number, corresponding to Eq. (12)
t_D	V_d/q_F, residence time in the zone of maximum energy dissipation
t_F	V/q_F, circulation time
v	$4q/\Pi D^2$, superficial velocity of gas
v_L	Velocity of gas leaving the gas distributor
u	Velocity of liquid
u_{ax}	Axial velocity
u′	Velocity of turbulent fluctuation
v_S	Settling velocity of particles
V	Fill volume
V_D	Volume with maximum energy dissipation
V_I	$\Pi d^2 h/4$, Volume enclosed by the impeller
V_δ	Volume of boundary layer
w	Width of baffles
X	Cell concentration
x; y	Coordinate
z	Number of impeller blades
z_F	Number of circulations through the impeller zone
z_I	Number of impellers
z_L	Number of holes in the sparger
α	Blade inclination to the horizontal
δ	Thickness of boundary layer
ε	Local energy Dissipation
$\bar{\varepsilon}$	$P/\varrho V$, Mass related impeller power

$\bar{\varepsilon}_a$ $P_a/\varrho V$, Mass related gassing power
ε_m Maximum energy Dissipation
η_L $(\nu^3/\varepsilon)^{1/4}$, Kolmogorov's length scale of turbulence
κ Adiabatic coefficient
Λ Macroscale of turbulence
ν Kinematic viscosity of liquid
ν_G Kinematic viscosity of gas
ϱ Density of liquid
ϱ_{Go} Density of gas under environmental pressure
ϱ_P Density of particle
σ Diameter related strength of particle
τ Laminar shear stress
τ_t Turbulent stress
ω Angular velocity

1 Introduction

The stress acting on particles is of high importance for many technical processes. As well as dispersion processes in two-phase systems (liquid/liquid or gas/liquid), where the disintegration of particles is desirable, there are also a number of processes that may be adversely affected by particle disintegration. These include precipitation, agglomeration, crystallisation processes, and also bioconversion with immobilised enzymes and the fermentation of sensitive microorganisms and animal and plant cells.

Even in the case of standard reactors such as stirred tanks and bubble columns, lack of knowledge in this area limits our ability to use particle stress as a selection criterion. The reasons for this lack of knowledge are, on the one hand, that the velocity fields in the reactors, which would allow a certain prediction, can only be obtained by sophisticated measurements and measurement techniques, and on the other hand, the stress on particles becomes evident only as an integral result of a long term process.

Many experimental results have been published, which deal with shear stress in biological systems. Most of them use laminar flow systems such as viscosimeters, flow channels or flasks and very small agitated vessels which are not relevant to technical reactor systems with fully developed turbulent flow. On the other hand the geometric and technical parameters are often not sufficiently described. Therefore it is not possible to explain the complex mechanism of force in bioreactors only on the basis of existing results from biological systems.

For basic studies it is very advantageous to use suitable model particle systems which are much better reproducible and can be performed in a much shorter time. The best comprehension can be derived from studies under technical flow conditions in real bioreactors and partial comparison with experiences with biological cultures.

The aim of this report is to examine the principles of shear stress on particles that would allow the design of bioreactors for technical use, mainly stirred tanks, bubble columns and loop reactors.

2
Causes of Particle Stress

The stress acting on particles is due to a relative velocity between the particles and the fluid. If their mean velocities also differ, contact between the particles or between a particle and the tank wall or the impeller elements leads to impact stress. However, this impact stress is negligible if the density differences and the particle concentrations are low.

Hydrodynamic stress exists independently of this. In the case of laminar flow, the stress is given by Newton's law (1):

$$\tau = \eta \frac{du}{dx} \tag{1}$$

In most industrial reactors turbulent flow is present and laminar flow exists only in boundary layers, which as a rule are of subordinate importance. In this situation, the relative velocity that determines the stress acting on particles is the velocity of the turbulent fluctuations. If one follows the concept of the theory of isotropic turbulence, according to which the turbulence of flow is explained by eddies of different sizes and with an energy content that decreases with their size, particle disintegration must be determined mainly by eddies of a size comparable to that of the particles. Clearly, larger eddies are followed by the particles as a convective movement, whereas smaller eddies have too low energy intensity to be important. If the particles are generally smaller than the smallest turbulent eddies, the size of which is approximately three times Kolmogorov's microscale of turbulence $\eta_L=(\nu^3/\varepsilon)^{1/4}$, even the smallest eddies cause stress (see Liepe [1] and Fig. 2).

In the velocity field of the determining eddies, which is characterized by the turbulent fluctuation velocity $\sqrt{u'^2}$, the particles experience a dynamic stress according to the Reynolds stress Eq. (2):

$$\tau_t = \varrho \sqrt{u'^2} \tag{2}$$

As is confirmed by the results in Section 6, the experimental findings correspond more closely to this stress formula, even in the dissipation range, than to Newton's stress formula (1), which is often used because of the assumption of laminar-flow eddies in this region (see e.g. [51, 52, 77]).

The turbulent fluctuation velocity $\sqrt{u'^2}$ at a distance $\Delta r = d_P$ between two neighbouring points in the velocity field depends on the size of the eddies. Depending on the energy dissipation ε and the particle size d_P, the determining eddies are those from the inertia range ($d_P > 25\,\eta_L$) to the dissipation range ($d_P < 6\,\eta_L$). According to Kolmogorov [2], for isotropic turbulence the following relationships are found. These may be assumed in the case of fully developed turbulence of the basic flow for many industrial plants, with the exception of boundary-layer flows.

Inertial range:

$$(0{,}1 \ldots 0{,}2)\,\Lambda > d_p > 25\,\eta_L: \; \sqrt{u'^2} = 1{,}9\,(\varepsilon\, d_P)^{2/3} \tag{3}$$

Dissipation range:

$$d_p < 6\,\eta_L\colon \sqrt{u'^2} = 0{,}0676\, d_P^2 \frac{e}{\nu} \tag{4}$$

The following general expression of dimensionless shear stress (5) is derived from Eqs. (2–4):

$$\frac{\tau_t}{\varrho\sqrt{\nu\,\varepsilon}} = A\left(\frac{d_P}{\eta_L}\right)^a \tag{5}$$

Accordingly, with the given material system and energy dissipation ε, the particle stress depends only on the ratio of particle diameter d_P to Kolmogorov's length scale of turbulence η_L. The constant A and the exponent a assume very different values for the various ranges of microturbulence. Whereas according to Eqs. (3) and (4) $A = 1.9, a = 2/3$ in the inertial range and $A = 0.0676, a = 2$ in the dissipation range, there is a transitional range for $25\,\eta_L > d_P > 6\,\eta_L$ with constantly changing values for A and a (see Fig. 1). The dependency in the transition range can be approximated by the values $A \approx 0.22$ and $a \approx 4/3$.

The validity of Eqs. (3–5) are bond on the condition of fully developed turbulent flow which only exists if the macro turbulence is not influenced by the viscosity. This is the case if the macro turbulence is clearly separated from the dissipation range by the inertial range. This is given if the macro scale Λ is large in comparison to Kolmogorov's micro scale η_L. Liepe [1] and Möckel [24] found out by measurement of turbulence spectra's the following condition:

$$\frac{\Lambda}{\eta_L} = \left(\frac{\Lambda\,\varepsilon^{1/4}}{\nu^{3/4}}\right) > 125\,\ldots\,250 \tag{6}$$

Due to the non-uniform energy distribution in the reactors, the particles are subjected to different stresses during their circulation in the reactor. This is in

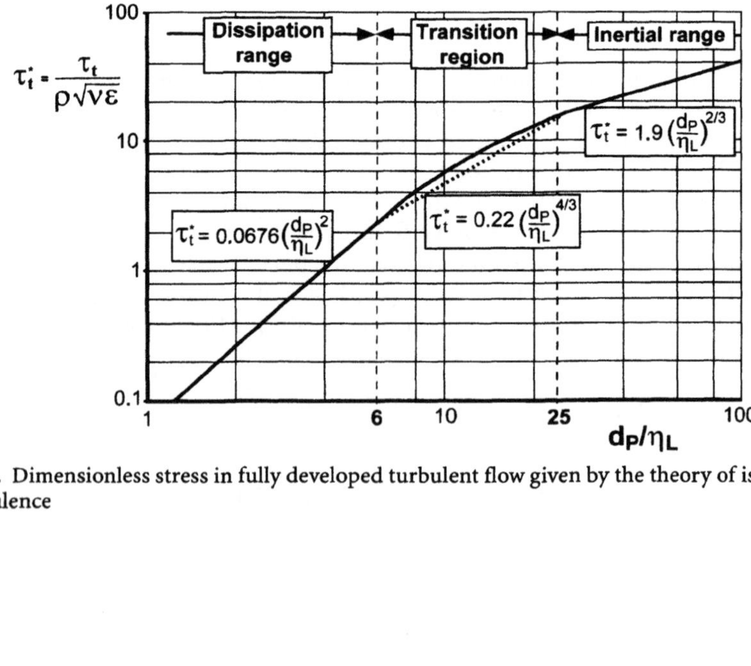

Fig. 1. Dimensionless stress in fully developed turbulent flow given by the theory of isotropic turbulence

addition to the variation of the local energy dissipation ε and therefore, as a result of the length scales Λ and η_L, also given by the reduction of particle size d_P in the case of a comminution process. In addition to the direct influence of ε and d_P as given by Eqs. (3) and (4) or the dependency in Fig. 1, it can also lead to a switch between the stress ranges.

The maximum energy dissipation and the location of its occurrence is generally not exactly known. Turbulence measurements provide some knowledge concerning non-uniform distribution, but give no comprehensive picture, since measurements are often very difficult or even impossible in the zones in which the power density is highest. This is true e.g. for the wake regions behind impeller blades in stirred tanks or for the gas/liquid boundary layers in gas-liquid contacting reactors.

Particle stress experiments are therefore particularly important since reliable conclusions regarding the maximum stress intensity in reactors of different types can be drawn from the overall result of the disintegration process.

3 Description of Reactors

In addition to true bioreactors which are used to carry out biological reactions, many authors have used special model apparatuses to study the stress on cells and organisms, and also non-biological particles, in models.

3.1 Bioreactors

Special reactors are required to conduct biochemical reactions for the transformation and production of chemical and biological substances involving the use of biocatalysts (enzymes, immobilised enzymes, microorganisms, plant and animal cells). These bioreactors have to be designed so that the enzymes or living organisms can be used under defined, optimal conditions. The bioreactors which are mainly used on laboratory scale and industrially are roller bottles, shake flasks, stirred tanks and bubble columns (see Table 1).

Whereas roller bottles and shake flasks are used for screening tests or for the cultivation of precultures, on production scale mainly stirred tanks, bubble columns, and in a few cases, also loop reactors are used (see Table 5). The stress in these reactors is therefore also of special significance and should be given particular consideration.

As the majority of technically important bioprocesses for the obtainment of useful materials and active substances are aerobic processes, the oxygen supply and CO_2 removal play an important part in the design and operation of bioreactors.

3.2 Model Reactors

Model reactors, e.g. viscosimeters (cylinder or cone-plate systems), channel currents and jets (see Table 2) have been used very often to test the shear stress

Table 1. Bioreactors

Reactors	Geometry
Shake Flasks	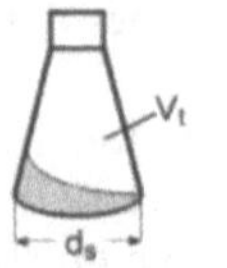
Stirred Tanks	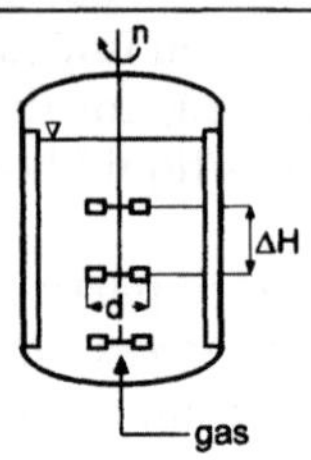
Bubble Columns	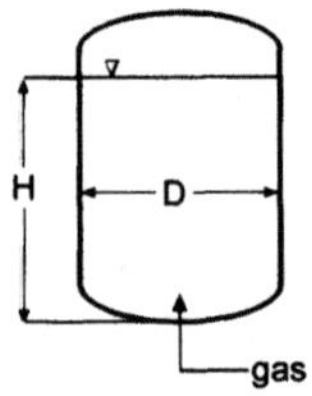

Table 2. Model reactors

Reactors	Geometry
Viscosimeters Searle: $\omega_2 = 0$ Couette: $\omega_1 = 0$	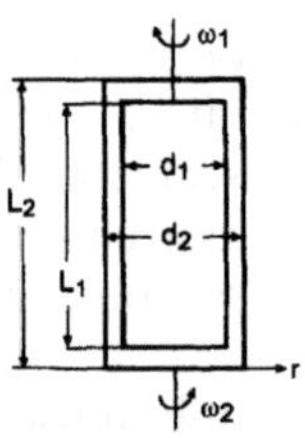 $d_2 \approx d_1$ $L_i \gg d_i$
Channel	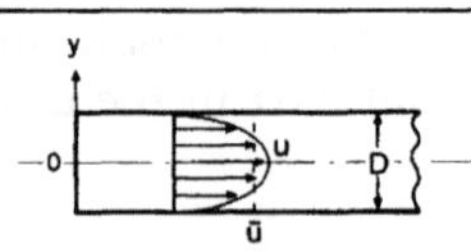
Jet	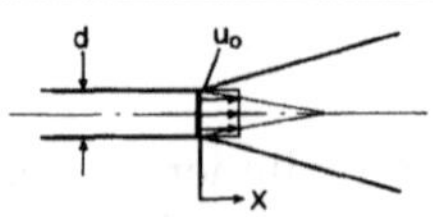

on particles, proteins, cells and organisms (see e.g. viscosimeters: [3–7], channel currents: [8–11] and jets: [12–13]).

In contrast to bioreactors, viscosimeters at laminar flow conditions have the advantage of relatively uniform and defined flow conditions, which permits the calculation of a certain shear stress. This is not the case for the other model reactors.

The general disadvantage of model reactors is that their flow conditions do not apply for bioreactors and therefore the transfer of stress experiments to technical applications is difficult or frequently impossible.

4 Estimation of Particle Stress on the Basis of the Velocity and Turbulence Fields

Stress on particles occurs in the velocity and turbulence fields of reactors. Therefore, for the initial estimate of stress according to Eqs. (1) and (2), the theoretically derived basic Eqs. of velocity fields can be used in the case of laminar flow, and the results of turbulence measurements in the case of turbulent flow.

In order to use Eqs. (3) and (4) or the data given in Fig. 1, for the calculation of maximum turbulent fluctuation velocity the maximum energy dissipation ε_m must be known. With fully developed turbulence and defined reactor geometry, this is a fixed value and directly proportional to the mean mass-related power input $\bar{\varepsilon} = P/\varrho V$, so that the ratio $\varepsilon_m/\bar{\varepsilon}$ can be described as an exclusive function of reactor geometry. In the following, therefore details will be provided on the calculation of power P and where available the geometric function $\varepsilon_m/\bar{\varepsilon}$.

4.1 Bioreactors

To avoid gas-liquid mass transfer limitation, which would have a negative influence on productivity, in correctly operated bioreactors there are turbulent flow conditions with more or less pronounced turbulence, for which the Reynolds stress formula (Eq. (2)) can be used. Whereas, as a rule there is fully developed turbulent flow in technical apparatuses (see condition (6) and explanations in Sect. 8), this is frequently not the case in laboratory fermenters. Equations (3) and (4) are then only valid to a limited extent.

4.1.1 *Shake Flasks*

In shake flasks there is neither undisturbed laminar flow nor fully developed turbulent flow. However, stress can be estimated approximately using Eqs. (2–4).

There is only very few process design literature for the calculation of power input for shake flasks [14–17]. Only the recent publication of Büchs et al. [17] provides a suitable correlation for the operating and geometric conditions existing for the practice. According to this, the following realationship applies to the technically important shake flasks without baffles:

$$Ne_s = \frac{P}{\varrho n^3 d_S^5} = \frac{1.94}{Re^{0.2}} \frac{V_L^{1/3}}{d_S} \quad \text{with} \quad Re = \frac{nd_S^2}{\nu} \tag{7}$$

This correlation was obtained as a result of extensive measurements with shake flasks of volume $V_t = 100-2000$ ml and corresponding inner diameter $d_S = 6.1-16$ cm, a filled volume of $V_L = (0.04-0.2)\,V_t$ and eccentricity of shaking machine of 25 and 50 mm.

Zoels [19] provides some results for shake flasks with baffles, which according to Henzler, Schedel [18] are not a better alternative than systems without reinforcement.

No details of energy distribution in shake flasks based on flow measurement are known to date.

4.1.2
Stirred Tanks

The power input in stirred tanks can be calculated using the equation: $P = Ne\,\varrho n^3 d^5$ if the Newton number Ne, which at present still has to be determined by empirical means, is known. For stirred vessels with full reinforcement (baffles, coils, see e.g. [20]), the only bioreactors of interest, this is a constant in the turbulent flow range: $Re = nd^2/\nu > 5000-10000$, and in the non-aerated condition depends only on geometry (see e.g. [20]). In the aerated condition the Newton number is also influenced by the Froude number $Fr = n^2d/g$ and the gas throughput number $Q = q/nd^3$ (see e.g. [21–23]).

It is known from many publications that aeration reduces the Ne number as a result of the formation of gas cavties behind the stirrer blades. The measured values can be described with sufficient accuracy using the following graphic correlation [23]:

$$\frac{Ne}{Ne_o} = \frac{1 - Ne_\infty/Ne_o}{1 + A\,Fr^a\,Q^b} = \frac{Ne_\infty}{Ne_o} \quad \text{with} \quad Ne_o \text{ and } Ne_\infty = f(Fr) \tag{8}$$

Here the Ne_o value is the Ne number in the condition without external aeration. It includes the reduction of power input which occurs as a result of surface aeration. The Ne_∞ value is the constant final value which is achieved with very high gas throughput numbers Q. Both values Ne_o, Ne_∞, the constant A and the exponents a, b are functions of the stirrer system geometry. In multi-stage stirred reactors, which are often used for fermentation, in addition to stirrer geometry, Ne_o and Ne_∞ for $H/(z_I\,d) < 2.5$ are also dependent to a large extent on average stage disdance $H/z_I d$ of impellers (see [23]).

According to current turbulence measurements in stirred tanks, there is a very considerable difference between the maximum energy density and the mean energy dissipation. Various authors (see e.g. [24]–[26] conclude that the ratio of the maximum energy dissipation to the mean energy dissipation $\varepsilon_m/\bar{\varepsilon}$ is very largely determined by the ratio of the impeller diameter to the tank diameter, d/D:

$$\frac{\varepsilon_m}{\bar{\varepsilon}} = \frac{A}{(d/D)^3} = \frac{B}{V_I/V} \tag{9}$$

This relationship is at the same relative height of the impeller blade, h/d, and the relative depth of the liquid, H/D, equal to the ratio of the filled volume $V = \Pi D^2 H/4$ to the impeller volume $V_I = \Pi d^2 h/4$, which could be more generally applicable.

For the standard disk turbine with h/d = 0,2 and H/D = 1, to which most of experimental data relate, the average value $A \approx 0.4$ can be derived from the publications mentioned.

Due to the liquid circulation in stirred tanks which transports all particles with a certain frequency through the impeller zone, they undergo the maximum shear stress.

As shown in Sect. 6.3.1 the dependence on geometry of Eq. (9) cannot fully explain the results for particle stress [27].

4.1.3
Gas-Liquid Contacting Reactors

As is the case with pure bubble columns and gas-operated loop reactors, most bioreactors in technical use are aerated with oxygen or air. Reactors with pure surface aeration, such as roller bottles, shake flasks and small stirred reactors or special reactors with membrane aeration, are exceptions. The latter are used for the cultivation of cells and organisms which are particularly sensitive to shearing (see e.g. [28–29]). The influence of gas bubbles in increasing stress has been described in many publications (see e.g. [4, 27, 29, 30]). In principle it can be caused by the following processes:

- Formation of gas bubbles at the gas distributor
- Coalescence processes between the gas bubbles
- Stress in the wake of bubbles as they rise
- Bubbles bursting on the surface of the liquid

Particularly high stress occurs when bubbles burst on the surface of the liquid, whereby droplets are eruptive torn out of the surface [32–36]. According to theoretical calculations, maximum energy densities occur in the region of the boundary surface shortly before the droplets separate [36]. The results calculated by Boulton-Stone and Blake [34] show that these are exponentially dependent on bubble diameter d_B. Whereas these authors found values of $\varepsilon \approx 10^5\,m^2/s^3$ with $d_B = 0.5$ mm, these are only $\varepsilon \approx 1\,m^2/s^3$ with $d_B = 5$ mm. The situation may be different regarding the droplet volume separated from the surface by the gas throughput and thus the number of particles which are exposed to high stress. The maximum for this value occurs with a bubble diameter of $d_B \approx 4$ mm (see [34]), and it is therefore feasible that there could be an optimal bubble size.

This is particularly valid when the necessity for bioreactors of transporting a certain amount of oxygen or CO_2 is taken into consideration (see Fig. 26).

The results provided in the literature for stress with biological particle systems, whereby gas distributors with small hole diameters, i.e. with smaller bubble sizes, have a more negative effect on cells (see e.g. [4, 30, 31]), are frequently not comparable, as in these studies there was differing stress during bubble formation at the gas distributor due to different hole velocities.

A measure of the energy which dissipates in the bubble column can be derived from the adiabatic compressor power:

$$\bar{\varepsilon}_a = \frac{P_a}{\varrho V} = \frac{\kappa}{\kappa - 1} p_o q_o \left[\left(\frac{p_l}{p_o}\right)^{\frac{\kappa-1}{\kappa}} - 1\right] \quad \text{with } p_l = p_o + \varrho g H + \varsigma \varrho_{Go} v_L^2/2 \tag{10}$$

$\bar{\varepsilon}_a$ in Eq. (10) is the mean energy dissipation. As the pressure p_l accounts for both hydrostatic pressure and the pressure loss in the gas distributor, it registers also the energy input for the gas distribution.

As a simplification, the term in Eq. (10) that accounts for the kinetic energy of the gas jets emerging from the gas distributor is based on the expression $\varsigma \varrho_{Go} v_L^2/2$, which is valid for incompressible flow. Experimental investigations show [27], that for relatively low gas velocities it is possible to represent the empirically determined loss coefficients ς as accurately with this simplification as by the use of expressions for compressible flow.

As the effect of aeration can still be predicted with less certainty than the stress caused by single-phase currents, due to the many influencing factors, experiments to investigate particle stress are particularly important.

4.2 Model Reactors

4.2.1 Viscosimeters

There is an analytical solution of the Navier-Stokes equations for the flow between two rotating cylinders with laminar flow (see e.g. [37]). The following equation applies for the velocity gradient in the annular gap in the general case of rotation of the outer cylinder (index 2) and the inner cylinder (index 1):

$$\frac{du}{dr} = 2 \frac{(r_2/r)^2}{(r_2/r_1)^2 - 1} (\omega_2 - \omega_1) \quad \text{for} \quad L_i \gg r_i \tag{11}$$

In the Couette viscosimeter the outer cylinder moves and the inner one is fixed ($r = r_2, \omega_1 = 0$), in the Searle viscosimeter is it vice versa ($r = r_1, \omega_2 = 0$).

Using Eqs. (11) and (2) the stirrer performance and the resultant Newton number Ne* for the laminar flow range can be derived with the formula $P = 2\pi L_i r_i^2 \omega_i \tau_i$:

$$Ne_i^* = \frac{P_i}{\pi \varrho L_i r_1^2 r_2^2 \omega_i^3} = \frac{4}{Ta_i} \quad \text{with } Ta = \frac{(r_2^2 - r_1^2)\,\omega_i}{\nu} \tag{12}$$

According to Eq. (11) the stress in the cylinder gap changes with the radius ratio (the maximum differences in shear stress are $\tau_1/\tau_2 = (du/dr)_1 / (du/dr)_2 = (r_2/r_1)^2$), so that there are only nearly equal stress conditions with small radius ratios.

Furthermore, in Searle cylinder systems, secondary currents occur above certain Taylor numbers, resulting in greater power input than that resulting with Eq. (12). Comparison with experimentally determined Ne* numbers shows that the laminar flow range in internally driven cylindrical stirrers in the range of radius ratios $r_2/r_1 = 1.05-2$ is only valid for Taylor numbers $Ta_{krit} = (r_2^2 - r_1^2)\,\omega_1/\nu < 400-200$ [38]. Above $Ta > Ta_{krit}$ secondary currents in the form of paired eddies, so-called Taylor eddies, develop. So-called laminar cellular flow is present initially, and as the Taylor number increases further, this gives way to turbulent cellular flow, and finally to turbulent flow. Small gap widths and the superposition of an axial flow stabilises the laminar flow for $Re_{ax} = u_{ax}\,(r_2 - r_1)/\nu > 40$, so that the critical Taylor number assumes higher values (see [38]).

In Couette cylinder systems, laminar flows extends to much higher Re or Ta numbers. The transition point for all radius ratios is above $Re = \omega_2 r_2 (r_2 - r_1)/\nu \geq 2000$ [39].

If there is no laminar viscosimeter flow, only the shear stress acting on the rotating cylinder surfaces can be calculated. It can be derived by the equilibrium of forces on the rotating cylinder:

$$\tau_{1/2} = \frac{P_{1/2}}{2\pi L_{1/2} r_{1/2}^2 \omega_{1/2}} = Ne_{1/2}^* \frac{\varrho r_{1/2}^2 \omega_{1/2}^2}{2} \tag{13}$$

τ_1 and Ne_1 in Eq. (13) are valid for the case of the Searle type and τ_2 and Ne_2 for the Couette type. The shear stress from Eq. (13) is the maximum shear which occurs in the gap close to the rotating cylinder. The uniformity of stress inside the gap decrease with increasing Re number. If the particles have the tendency to flow close to the moving wall, they will be subjected to the maximum shear.

4.2.2
Channels

The special flow conditions in circular (capillaries, tubes) or rectangular channels cause very different stresses depending on the position of the particles in the flow cross section. With laminar flow, for example the following applies to velocity gradient (see e.g. [37]):

$$\frac{du}{dy} = C \frac{\bar{u}}{D} \left(\frac{2y}{D} \right) \tag{14}$$

where y is the position coordinate which starts in the centre of the flow channel.

$C = 8$ applies to circular cross sections. For rectangular channels with large width-to-height ratios $b/D \gg 1$, $C = 6$. Equation (14) is valid for pipe flow for $Re = \bar{u}\,D/\nu < 2300$, the transition point for rectangular channels is at $Re \approx 1500$.

According to Eq. (14) the maximum velocity gradient at the wall is at $y = D/2$. It amounts to $(du/dy)_{wall} = C\,\bar{u}/D$. The stress derived from this with Eq. (1) has been used in a number of studies (see e.g. [8–11]) as a measure of stress. However, particles are only subjected to this maximum stress if they close to $y = 0$. As this cannot or can only temporarily be the case during flow through channels, such test results should be regarded with caution and only conditionally suitable for comparison with the results from other model apparatuses, not to mention bioreactors.

With turbulent channel flow the shear rate near the wall is even higher than with laminar flow. Thus, for example, $(du/dy)_{wall} = 0.0395\ Re^{3/4}\,\bar{u}/D$ is valid for turbulent pipe flow with a hydraulically smooth wall. The conditions in this case are even less favourable for uniform stress on particles, as the layer flowing near the wall (boundary layer thickness δ), in which a substantial change in velocity occurs, decreases with increasing Reynolds number according to $\delta/D = 25\ Re^{-7/8}$, and is very small. Considering that the channel has to be large in comparison with the particles: $D \gg d_P$, so that there is no interference with flow, e.g. at $Re = 2300$ and $D = 10\ d_P$ the related boundary layer thickness becomes only approx. 29% of the particle diameter. It shows that even at $Re = 2300$ no defined stress can be exerted and therefore channels are not suitable model reactors.

4.2.3
Jets

The flow of jets becomes turbulent at much lower Re numbers than channel flows. Calculating the stress from the mean velocity profiles does not reflect the true situation in turbulent flow. As in the case in most bioreactors, the maximum turbulent stress is determined by the turbulence, which can be calculated using Eqs. (2)–(4). It occurs in free jets after the nozzle, at the edge of the mixing zone. The following is generally valid:

$$\varepsilon_m = E\,\frac{u_0^3}{d} \tag{15}$$

In this equation u_0 is the nozzle exit velocity and d is the nozzle diameter.

According to turbulence measurements performed by Möckel [24] the maximum energy dissipation occurs in this zone with $E \approx 0.13$.

Baldyga et al [40, 41] derived energy density indirectly from experiments on the influence of micromixing on conversion in rapid chemical reactions. For the range of similarity of the free jets, they state the constant as function of length of coordinate x: $B = 50\ (x/d)^{-4}$ (see Table 2). At the beginning of the range of similarity at $x \approx 4.5\,d$ the result is $E \approx 0.12$, i.e. approximately the same maximum energy density as according to [24].

5
Methods for the Determination of Stress

In order to characterise living organisms in terms of permitted stress, these organisms must be tested under conditions which are the same as or as close as possible to practical conditions. As a rule this means that the type of reactor used for the technical application has to be used under comparable operating conditions. The majority of studies described in the literature do not fulfil this condition, so that their results are only conditionally applicable. Either idealised apparatuses (see Sect. 3.2) were used or the bioreactors were too small too accommodate the flow conditions with fully developed turbulence (see Sect. 2) corresponding to the technical scale. Furthermore, the important condition of sufficient and comparable mass transport (O_2 supply and CO_2 desorption) has not been considered or mentioned in many studies.

The use of suitable model particle systems is recommended for the comparative test of bioreactors and their operating conditions. They permit faster, more reproducible and thus more cost-effective optimisation of technical relevant reactors [27, 42–52].

5.1 Investigations with Living Organisms

If model reactors (e.g. viscosimeters, channels or jets) are used for the test, as there is no oxygen supply or CO_2 desorption, these have to be operated in the bypass of the bioreactor or by sampling from a bioreactor. This frequently causes additional stress as a result of the pumping, flow through connecting pipes and cross section changes at connecting pieces, sampling, transport and collecting biosuspension, which must be kept small in comparison to the effects of the model reactor.

The following methods are usually used to determine stress:

- Image analysis in case of size change in organic material, pellets, mycelium hyphae or agglomerates of single cells (see e.g. [44–54, 60])
- Cell count: staining method with Trypan Blue (see e.g. [68])
- Cell mass determination
- Determination of product concentration
- Analysis of the intracellular enzyme, lactate dehydrogenase (LDH, see e.g. [68]), in the supernatant
- Protein analysis for enzymes (see e.g. [44, 47, 49])

5.2 Model Particle Systems

Model particle systems have to have the following properties:

- Small density difference from the liquid used
- Viscosity similar to the biological broth
- Particle size and sensitivity to stress similar to biological systems
- Stress should be easily and reproducibly measurable

Since in the case of turbulent stress the ratio of particle diameter d_P to length scale of turbulence η_L is decisive for the stress regime (see Fig. 1) the model particle systems must have properties which guarantee d_P/η_L values which are in the same range as for the biological particle systems.

Table 3. Model particle systems

Material system	C [g/l]	d_p [µm]	ϱ_P [g/l]	v_S [m/h]	Measured value	Registration
Blue clay	5					Laser scanning
Prästol PR 650 BC	0.01	500 –	1140 –	22 – 3	Particle	microscope
NaCl	1	10	2600		diameter	PAR-TEC 100
water 1000 µS/cm	–					(Image analysis)
Enzyme resin:						
Pen G-Acylase immobilized on Ion exchanger OC 1050	60	240 – 150	1080	8 – 3	Enzyme activity	Enzyme test
Phosphate puffer pH = 8.1	–	–	–	–		
Silicon oil PH 300	2					
Emulsifier	0.7	500 –	1060	22 –	Particle	Image analysis
Azo-dye Ceres red 32	0.001	20		0.05	diameter	

Recommended model particle systems are enzymes immobilised on carriers ([27, 44, 45, 47, 49]), oil/water/surfactant or solvent/water/surfactant emulsions ([27, 44, 45] or [71, 72]) and a certain clay/polymer floccular system ([27, 42–52]), which have proved suitable in numerous tests. The enzyme resin described in [27, 44, 47] (acylase immobilised on an ion-exchanger) is used on an industrial scale for the cleavage of Penicillin G and is therefore also a biological material system. In Table 3 are given some data to model particle systems.

As a measure of particle stress the particle diameter, or the enzyme activity in the case of immobilised enzymes, can be used.

Due to the stress the size of particles decreases. Since the recommended particle systems show only a very small tendency if any to re-agglomerate outside the high stress region, the particle sizes should be generally reduced until an equilibrium particle size is reached after prolonged exposure to the stress. This equilibrium particle size is determined by the balance of forces between the strength of the particle τ_P and the stress τ acting on the particle. If the particle strength is known, therefore, the stress acting on the particle can be deduced from the relationships found experimentally for the equilibrium particle size. Since the strength for the floccular and emulsion systems follows the dependency $\tau_P = \sigma/d_P$ (see [27]) the Reynolds stress formula (2) gives the following relationships for the two regions, Eqs. (3) and (4) of the turbulence spectrum:

Inertial range:

$$d_p \approx 0.68 \left[\frac{\sigma}{\varrho\, \varepsilon_m^{2/3}} \right]^{0.6} \tag{16a}$$

Dissipation range:

$$d_p \approx 2.46 \left[\frac{\sigma \nu}{\varrho\, \varepsilon_m} \right]^{1/3} \tag{16b}$$

Since in the case of using immobilised enzymes (acylase enzyme resin [27, 45, 47]), normally only a small decrease in the particle diameter leads to a distinct decrease in the enzyme activity, the more sensitive enzyme activity has to be taken as the measure of stress.

The equilibrium particle diameter in the case of non agglomerate particle systems or the enzyme activity of immobilised enzymes after a certain exposure of time is entirely due to the reactor-specific comminution process, and conclusions can therefore be drawn regarding the maximum intensity of hydrodynamic stress.

The following methods can be recommended to determine the particle diameter

- Microscopically by image analysis
- Laser scanning microscope in the case of the floccular system, whose signals were calibrated by the absolute microscopically measurements mentioned [45]. The sensor of this instrument, which is a rotating laser beam, is located in a measurement probe, which allows direct measurement within the flow of the reactor. Together with the relatively short disintegration times for the floccular system, this possibility of in-situ measurement and automatic recording makes the measurement of the disintegration process particularly reproducible, easy and fast.

The comparison of biological material systems and model particle systems in Fig. 2 shows that, under the operating conditions relevant for bioreactors, the

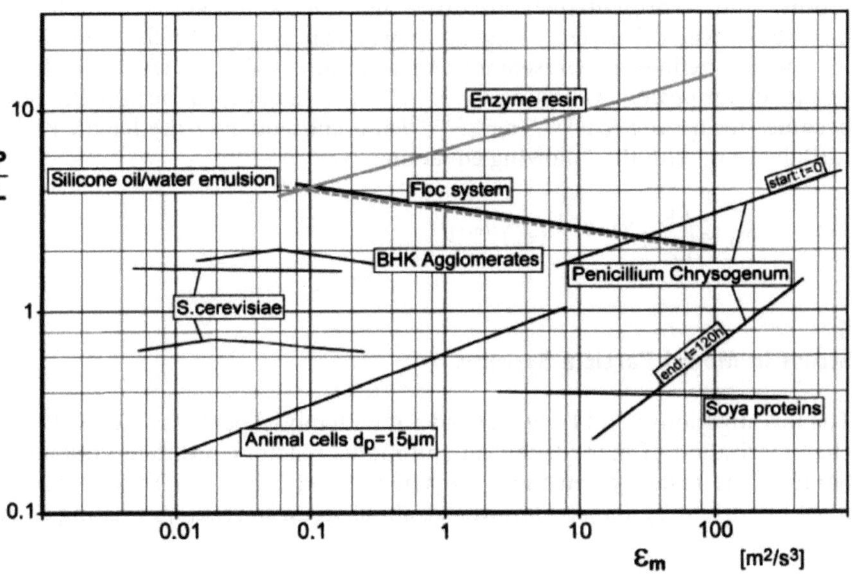

Fig. 2. Properties of model and biological particle systems: Micro scale related particle diameter d_P/η_L versus maximum energy dissipation ε_m in stirred reactors explanations see Table 3 and Table 4

Table 4. Conditions for the investigations with biological particle systems

Particle system	X [cells/ml]	V [L]	Impeller	$\frac{d}{D}$	$\frac{\varepsilon_m}{\bar{\varepsilon}}$	Ref.
S. cerevisae	$5 \cdot 10^6$ $4.9 \cdot 10^7$	1.5	turbine	0.36	31	[57]
BHK-cells	$2 \cdot 10^7$	0.25	blade	0.72	35.5	[58]
protein precipitate	–	14.5	pitched blade	0.41	61	[59]
Penicillium chrysogenum	–	6	blade turbine pitched blade	0.6 0.33 0.4	6.5 36 124	[60]

enzyme resin for $\varepsilon_m < 1\ m^2/s^3$ and the other particle systems recommended in Table 3 for the hole range of technical working conditions are sensitive to stress in a similar range of microturbulence $d_P/\eta_L < 6$ as many biological material systems (agglomerates of BHK cells as well as yeasts, mycelium hyphae of Penicillium chrysogenum, single animal cells and also soya protein agglomerates).

The reactor conditions for the use of biological material systems from [57–59], which are stated in Fig. 2 are provided in Table 4. In the case of Penicillium chrysogenum, where there are non-Newtonian fermentation broths, the representative viscosity in the Kolmogorov length scale $\eta_L = (\nu^3/\varepsilon)^{1/4}$ was calculated using the power concept of Henzler and Kauling [61], which defines a representative shear gradient as $\gamma = \sqrt{\varepsilon/\nu}$. With this dependency, for power law liquids: $\eta = K \cdot \gamma^{m-1}$ the viscosity in the impeller region becomes to: $\eta = \varrho (K/\varrho)^{2/(m+1)}\ \varepsilon_m^{(m-1)/(m+1)}$ and subsequently the length scale of turbulence η_L with the following equation.

$$\eta_L = \left(\frac{K}{\varrho}\right)^{\left(\frac{3}{2(m+1)}\right)} (\varepsilon_m)^{\frac{m-1}{4(m+1)}} \tag{17}$$

6 Stress in Model Particle Systems

The investigations [27, 44–49] carried out with model particle sytems allow the characterisation of many technical and model reactors and their comparison. Some of the results given there are summarised here since they contain the most important, systematic knowledge about stress in reactors existing so far.

6.1 Investigated Systems

The results provided here were obtained with the reactors listed in Table 5.

Table 5. Special reactors used with model particle systems

Reactors		Geometry	
Stirred tanks		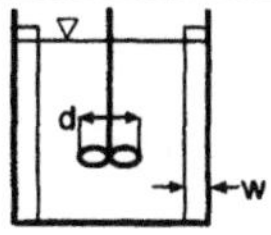	D = 0.15 – 1 m H/D = 1 … 2 d/D = 0.2 … 0.7 w/D = 0.1
with bondary layer flows	Stirred tanks without baffles	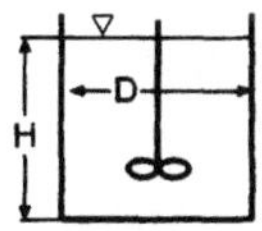	D = 0.15; 0.4 m H/D = 1 d/D = 0.2 – 0.7
	Stirred tanks with smooth disc	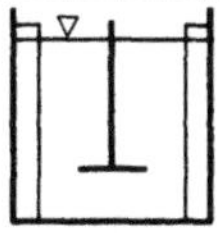	D = 0.4 – 1 m H/D = 1 d/D = 0.65 w/D = 0.1
Viscosimeter	Typ: Searle.	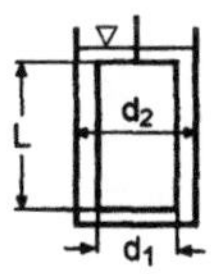	d_2 = 42.08 mm d_1 = 40 mm L = 60 mm
with gas-liquid bondary layers	Bubble column	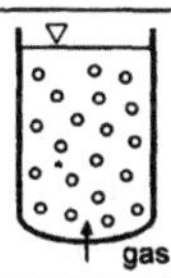	D = 0.4; 1 m H = 0.08 – 2.08 m
	Loop reactors	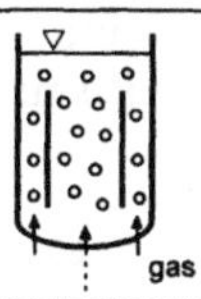	D = 0.4 H/D = 2

To determine the influence of stirrer type and geometry, as regards the use of stirred tanks, many different types of stirrers were tested (see Table 6).

In the case of studies in stirred reactors and bubble columns the vessel diameters were in the range of D = 0.15 to 1 m and the filling height in the range of H = 0.15 to 2.08 m. The larger dimensions were used to obtain technical relevance and reliable scale-up rules. Especially important are the dimensions of the turbulence producing element, which is in the order of the macroscale of turbulence. The macroscale has to be large in comparison with Kolmogorov's legth scale of turbulence, see Eq. (6) to achieve fully turbulent flow, as in technical reactors.

6.2 Destruction Kinetics

Figure 3 shows some examples of kinetic curves $d_F(t)$ for the disintegration of flocs in stirred tanks.

Table 6. Investigated impellers

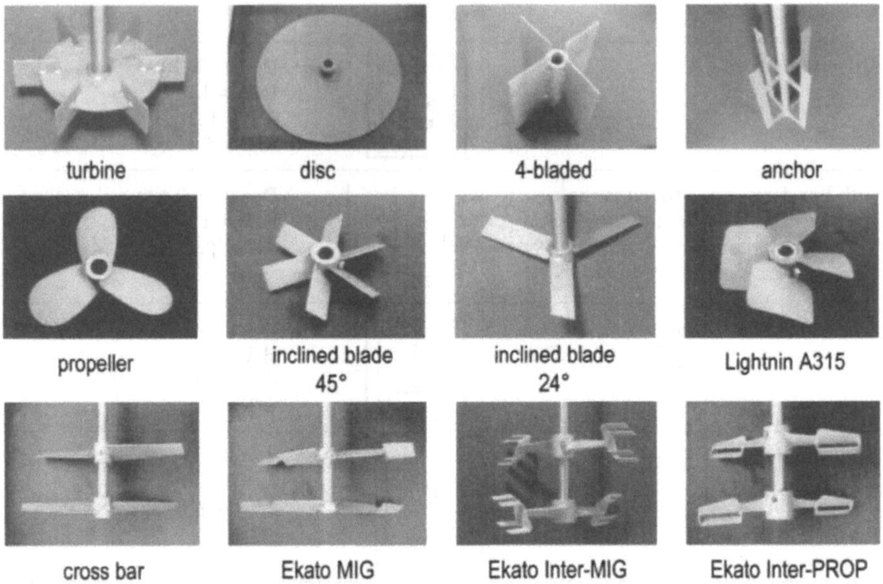

The strong influence of the specific impeller power P/V and of the impeller type on the disintegration of flocs can be seen from these curves, so that the effects of the operating conditions and the reactor type can be determined with satisfactory accuracy. The disintegration kinetics are complex, and show an exponential decrease in particle size with time.

$$\text{floc sytem} \qquad \frac{d_F - d_{F\infty}}{d_{F0} - d_{F\infty}} = \exp\left(-E_1 \bar{\varepsilon}^e\, t\right)$$

$$\text{enzym resign} \qquad \frac{a - a_\infty}{a_0 - a_\infty} = \exp\left(-E_2\, t\right) \qquad (18)$$

where for the floccular system $e \approx 0.5$ and $d_{F\infty} \approx d_{Fv}$ were found. Since the constants e, E_1, E_2 within the scope of measurement accuracy, proved not be dependent on agitator geometry, the destruction kinetics are not visible influenced by the circulation behaviour of the agitator. A similar course is found for the droplet diameter in the case of the oil/water emulsion. The droplet diameter tends towards an equilibrium value after $t \leq 6$ h, and this value can be used for comparisons. In the case of the floccular system, on the other hand, no real equilibrium state is reached even at the lowest stress, because of the ageing of the polymer [44, 47]; instead, the particles are reduced to the primary particle diameter of blue clay. For ease of comparison of the various reactors, therefore, a reference floc diameter d_{Fv} is defined for the floccular system as the diameter found at a disintegration rate of $d(d_F)/dt = 0.0055$ mm/s. This rate is low in com-

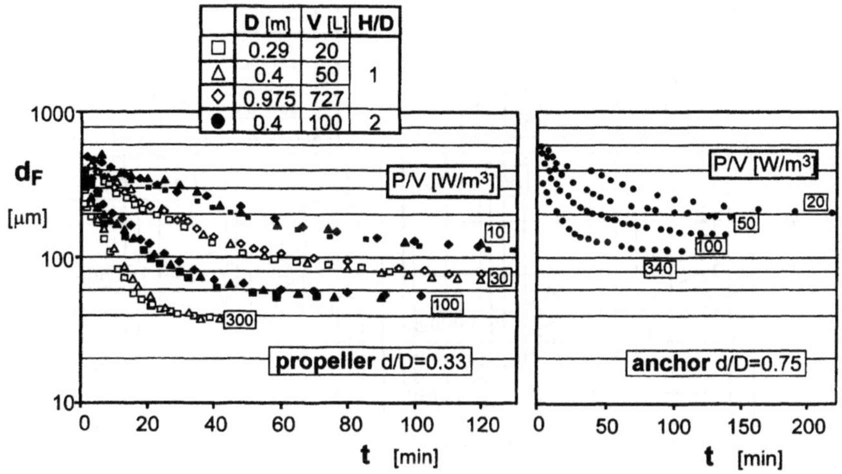

Fig. 3. Destruction kinetics of floc system for stirred reactors: 4 baffles; w/D = 0.1; H/D = 1; D = 0.4 m

parison with the initial change in the particle diameter and also in comparison with the mean change, and the reference diameter can be determined within an acceptable test time of $t < 2$ h even when the stresses are low. The reference floc diameter d_{Fv} has approximately the same value as the equilibrium floc $d_{F\infty}$ diameter, which can only be identified by means of Eq. (18).

In the case of the enzyme resin, the relative enzyme activity a/a_0 after $t = 300$ h was used as the reference parameter.

6.3 Floccular Systems

6.3.1 *Baffled Stirred Tanks*

As can be seen even from the kinetic curves in Fig. 3, the type of impeller has a decisive influence on particle disintegration in stirred tanks. This is particularly clear from a comparison of other impeller systems on the basis of the reference particle diameter d_{Fv} in Fig. 4.

Contrary to commonly held opinion (e.g. [62, 63]), this comparison shows that axial-flow impellers such as pitched-blade impellers and propellers lead to particularly high stresses at the same specific impeller power. The impeller geometry, such as the impeller-to-tank diameter ratio d/D and the relative blade height h/d (see Fig. 5), also has a distinct influence.

Since the results demonstrate the important role of the impeller size, the first approach adopted for the correlation of the test results was to try to use the generalised formula of Eq. (9) for the maximum energy dissipation on the basis of the turbulence measurements.

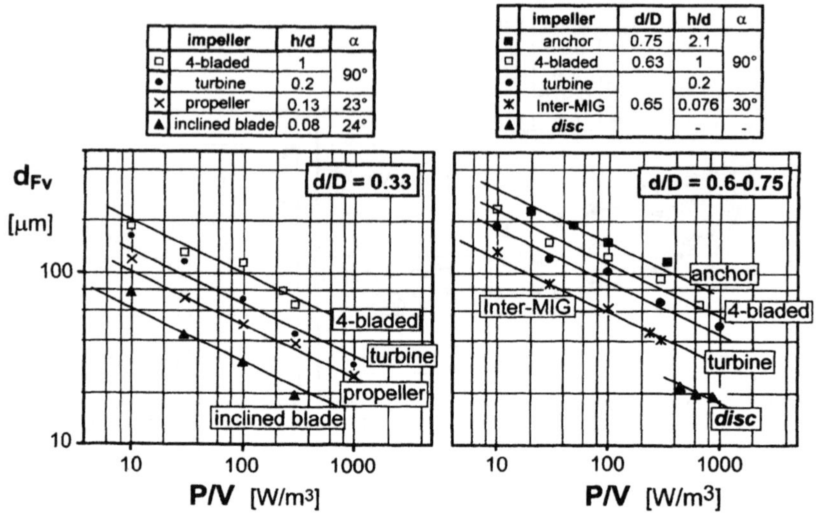

Fig. 4. Influence of impeller type on stress: Reference floc diameter d_{Fv} in dependency on specific impeller power P/V; 4 baffles; w/D = 0.1; H/D = 1; D = 0.4 m

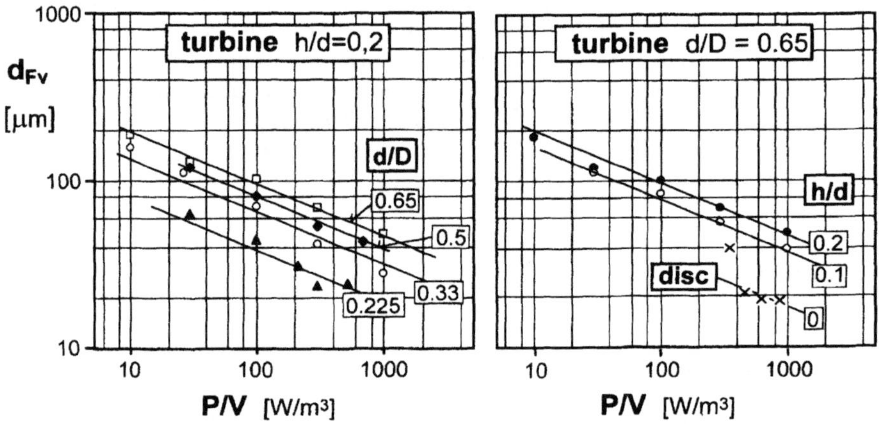

Fig. 5. Influence of diameter ratio d/D (left hand diagram) and diameter related blade h/d (right hand diagram) on stress: Reference floc diameter d_{Fv} in dependency on specific impeller power P/V; 4 baffles; w/D = 0.1; H/D = 1; D = 0.4 m

This formula also leads to good clustering of the results of the measurements for disc impellers (see Fig. 6) and for the other radial-flow impellers such as paddle and anchor impellers (see Fig. 7). Unsatisfactory correlation is found especially for axial-flow impellers, which show a systematic downward deviation in Fig. 7.

The other geometrical parameters such as the number of blades and the blade angle, which are not included in the ratio of the volume of fill to the impeller volume, evidently play a definite role (see Fig. 8).

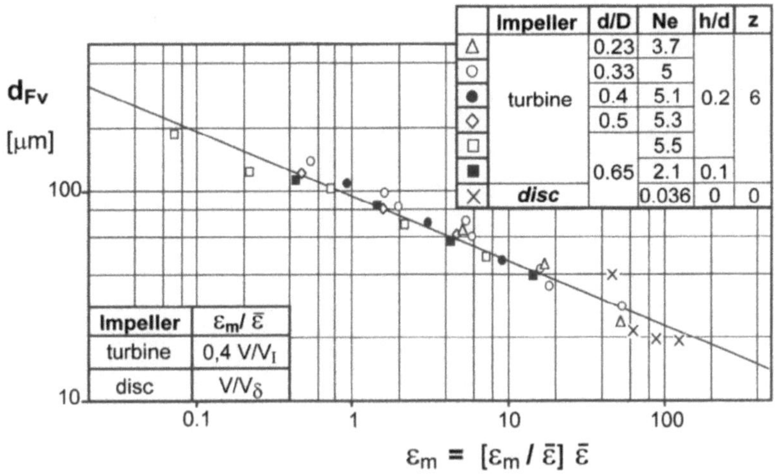

Fig. 6. Correlation of floc diameter d_{Fv} for turbine impeller and a smooth disk: 4 baffles; w/D = 0.1; H/D = 1; D = 0.4 m

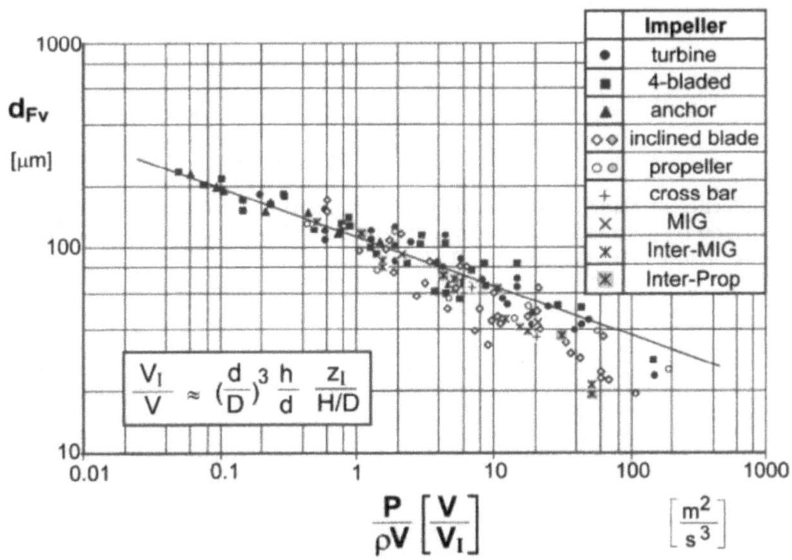

Fig. 7. Reference floc diameter d_{Fv} versus impeller power per impeller volume P/V_I for radial and axial impellers; 4 baffles; w/D = 0.1; H/D = 1; D = 0.4 m

The results in the upper diagram of Fig. 8 relate to special impellers having a constant ratio V_I/V, and so confirm only the influence of the blade angle. The results show in Fig. 8 that a smaller blade angle and also a smaller number of blades (lower diagram) lead to increased stress on the particles.

A regression formula that also takes these geometrical values into account is therefore used for better correlation. Very good clustering of all of the results can be achieved with the correlation function (19) (see Fig. 9).

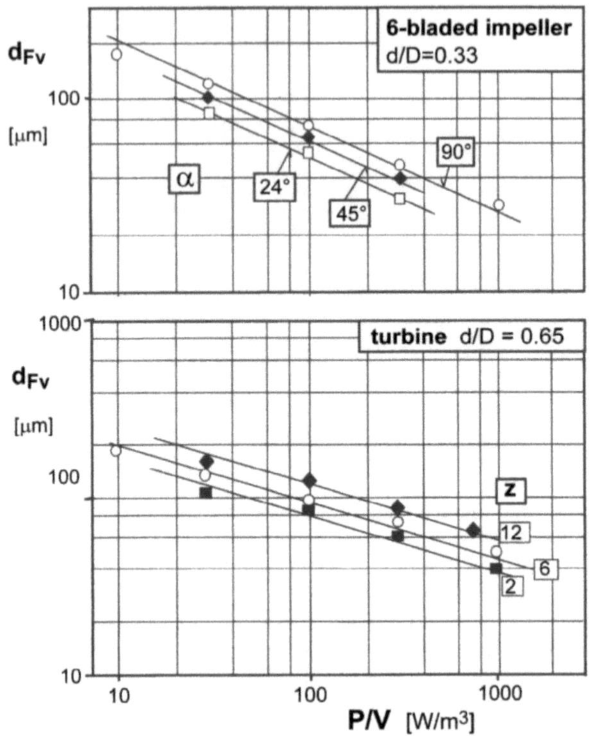

Fig. 8. Influence of blade inclination α (upper diagram) and number z of impeller blades (lower diagram) on stress: Reference floc diameter d_{Fv} in dependency on specific impeller power P/V; 4 Baffles; w/D = 0.1; H/D = 1; D = 0.4 m

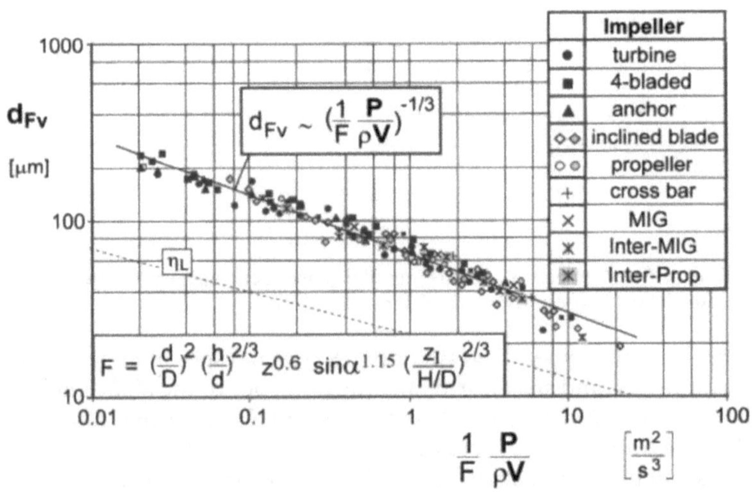

Fig. 9. Correlation of floc diameter d_{Fv} for radial and axial impellers: 4 baffles; w/D = 0.1; H/D = 1 – 2; D = 0.4 m

$$d_{Fv} = \frac{G}{\left(\frac{1}{F}\frac{P}{\varrho V}\right)^{1/3}} \tag{19}$$

with

$$F = \frac{P}{P_D}\frac{V_D}{V} = \left(\frac{d}{D}\right)^2 \left(\frac{h}{d}\right)^{2/3} z^{0.6} \sin\alpha^{0.6}\, z_I^{2/3} \left(\frac{H}{D}\right)^{-2/3}$$

Should the largest part of the power P_D dissipate into the region of the stirrer, making $P_D \approx P$ valid, the geometrical function F could then be regarded as the ratio of volume V_D in which the largest energy density occurs, to the total volume of the reactor V, analogous to the ratio V_I/V used in equation (9).

Where the Reynolds stress formula (2) and the universal law of the theory of isotropic turbulence apply to the turbulent velocity fluctuations (4), the relationship (20) for the description of the maximum energy dissipation can be derived from the correlation of the particle diameter (see Fig. 9). It includes the geometrical function F and thus provides a detailed description of the stirrer geometry in the investigated range of impeller and reactor geometry: $0.225 < d/D < 0.75$, $0.1 < h/d < 1$, $2 < z < 12$, $24° < \alpha < 90°$, $z_I = 1; 2$, $1 < H/D \leq 2$. This geometry function shows a different dependence of the ratios d/D and h/d from that derived from many turbulence measurements, correlation (9).

$$\frac{\varepsilon_m}{\bar{\varepsilon}} = \frac{a}{F} = \frac{a}{(d/D)^2 (h/d)^{2/3} z^{0.6} \sin\alpha^{1.15}\, z_I^{2/3} (H/D)^{-2/3}} \tag{20}$$

The comparison of the correlation functions of Fig. 6 and Fig. 9 for the disk turbines with analogues geometry produces an average approximately value of a ≈ 4 for the constant in Eq. (20).

The as yet unpublished results of more recent turbulence measurements confirm the geometrical dependence of Eq. (20) as regards the ratio h/D and blade number z for radial stirrers. For the influence of the diameter ratio d/D, the relationship approximating to Eq. (20) $\varepsilon_m/\bar{\varepsilon} \propto (D/d)^{2.35}$ was derived from these tests, in agreement with Zhou and Kesta [64]. However, turbulence measurements in axial stirrers resulted in substantially lower energy dissipation [64].

Therefore, it is yet to be clarified whether the description of the particle destruction process with Eqs. (2–4) or the simplification in the estimate of energy dissipation from the measured turbulent kinetic energy produces these differences.

As shown in Sects. 6.4 and 7, regardless of this, the energy dissipation according to Eq. (20) together with Eqs. (2–4) results in correct conclusions as regards particle stress for widely varying particle systems and all stirrer types tested here which can be used in practice of particle destruction.

6.3.2
Reactors with Dominating Boundary-Layer Flow

If boundary-layer flow plays an important role in reactors, as is the case e.g. in unbaffled stirred tanks or in agitation with a smooth disc (reactors see Table 5),

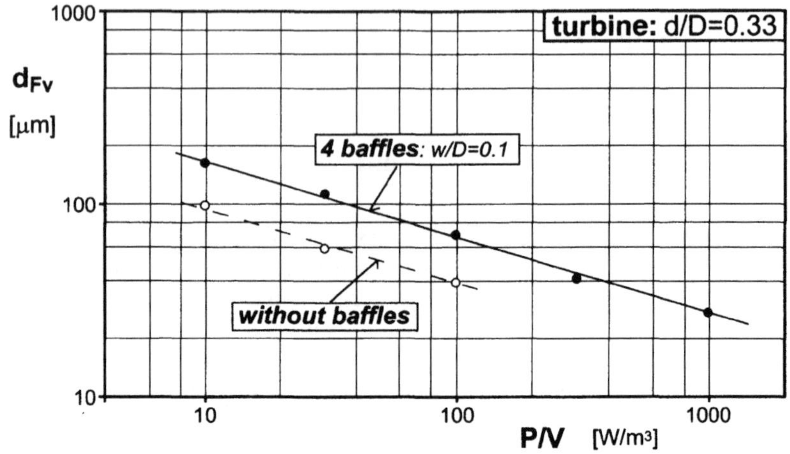

Fig. 10. Comparison of stirred vessels with and without baffles: Reference floc diameter d_{Fv} in dependency on specific impeller power P/V; H/D = 1; D = 0.4 m

the floc diameter is smaller than that found for the same average power input P/V with baffled impellers (see Fig. 10 and right-hand diagram in Fig. 5).

What happens is here that the flocs evidently move into the thin boundary layers adjacent to the hydraulically smooth surfaces, where much of the energy is dissipated, with the result that the particles are subjected to strong stresses because of the small volume of the boundary layers. This hypothesis is supported by the good correlation of the results for the smooth disc with the results for various impellers in Fig. 6; it was assumed here that in the case of the disc, the majority of the power is dissipated in the boundary-layer volume V_δ, and the relationship $\varepsilon_m/\bar{\varepsilon} \approx V/V_\delta$ is approximately valid. The volume of the boundary layer (Eq. (21)) was obtained by integration from the theoretical solution [65] for the thickness of the boundary layer (Eq. (21)) of a smooth disc with turbulent flow.

Boundary-layer thickness:

$$\delta(r) = 0.48 \frac{r}{Re^{1/5}} \quad \text{for} \quad Re = \frac{n d^2}{\nu} > 10^5 \tag{21}$$

Boundary-layer volume:

$$\frac{V_\delta}{V} = \frac{2\Pi}{V} \int_0^{d/2} r\,\delta(r)\,dr = 0.37 \frac{H/D}{(d/D)^3} \frac{1}{Re^{1/5}} \tag{22}$$

The factor 0.48 in Eq. (21) is valid for a rotating disc in infinite space. As is shown by a comparison of the Ne numbers found experimentally, $Ne = 0.52/Re^{0.2}$, with Schlichting's data [65], this condition is obviously satisfied very well for the design ratios chosen here ($h_B/d = 0.7$ with $H/D = 1$).

For the range involved here, which is $5 \times 10^5 < Re < 7 \times 10^5$, Eq. (22) gives an average boundary layer thickness of $\delta \approx 2.8$ mm, which is much greater than the reference floc diameter, so that disintegration of the flocs in the boundary layer also seems plausible. This is also true of the boundary layer thickness at the wall of the stirred tank, which is found to be $\delta_W \approx 0.9$ mm [27].

6.3.3 *Viscosimeters*

Viscosimeters were included in the investigations to provide a link with the many results in the literature for shearing experiments with biological cultures.

The results presented here were found by investigations with a special cylinder system [45, 48]. This system was constructed for an existing Searle viscosimeter (rotation of inner cylinder), such that the gap widths were large in relation to the reference floc diameter of the floccular system used, so that the formation of the flocs and their disintegration in the cylinder system are not impaired. For this system, with $r_2 = 22$ mm, $r_1 = 20.04$ mm, and $L_1 = 60$ mm ($r_2/r_1 \geqslant 1.098$), the following Newton number relationships were determined from the experimental values collected by Reiter [38] for the Taylor number range of $400 < Ta < 3000$ used here:

$$Ne^* \approx \frac{0.32}{\sqrt{Ta}} \quad \text{for} \quad 200 \leq Ta \leq 2 \times 10^4 \tag{23}$$

In these investigations, the viscosimeter and a stirred tank with laser probe [45, 48] were arranged in a pumped circulation system. The experimental set-up was designed in such a way that the movement caused by the impeller and the pumping produced negligible stress on the particles. For the resulting axial Reynolds number $Re_{ax} = u_{ax}(r_a - r_i)/\nu = 21-38$, according to [38] laminar cellular flow should be present in the entire range of $800 < Ta < 6000$.

Figure 11 shows the reference floc diameter for viscometers as a function of shear stress and also the comparison with the results for stirred tanks. The stress was determined in the case of viscosimeters from Eq. (13) and impeller systems from Eqs. (2) and (4) using the maximum energy density according to Eq. (20). For $\tau > 1$ N/m^2 ($Ta > 2000$), the disintegration performance produced by the flow in the viscosimeter with laminar flow of Taylor eddies is less than that in the turbulent flow of stirred tanks. Whereas in the stirred tank according to Eq. (4) and (16b) the particle diameter is inversely affected by the turbulent stress: $d_P \sim 1/\tau_t$, in viscosimeters it was found for $\tau > 1.5$ N/m^2, independently of the type (Searle or Couette), the dependency: $d_P \sim 1/\sqrt{\tau}$ (see Fig. 11).

The finding of greater stresses for $\tau < 0.8$ N/m^2 with the Searle viscosimeter flow than in stirred vessels may caused in a non uniform energy distribution in the viscosimeter also. In the boundary layer on the rotating inner cylinder high shearing stresses are present. At lower rotation speeds, the flocs also pass into these regions. At higher rotation speeds, on the one hand, the boundary layer thickness diminishes [27], and on the other hand, because of the density difference between flocs and solution increasing centrifugal effects (force action $\propto (\varrho_P - \varrho)\, r_1 \omega_1^2$)) are active which lead to a concentration of the particles outside the boundary layer on the inner wall so that the flocs are less stressed than in stirred vessels.

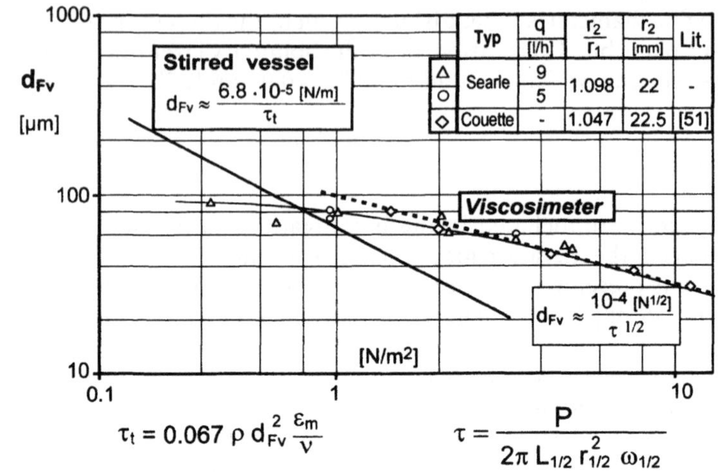

Fig. 11. Comparison of viscosimeter and stirred vessels: Reference floc diameter d_{Fv} in dependency on stress (stirred vessels: τ_t, viscosimeter: τ)

From these results it seems to be obvious that experimental investigations in viscosimeters could not characterise the relationships in the turbulent flow of stirred tanks.

6.3.4
Bubble Columns

As in boundary-layer flows, smaller reference floc diameters are found with gas sparging than with the same average power input in a baffled stirred tank [27] or [44, 45]. This can be explained if it is assumed that the flocs come into close contact with the gas phase and find their way into the zones of higher stress.

The effects of the different stresses mentioned in Sect. 4.1.3 cannot be determined individually by experimental studies, so that only collective conclusions are possible. Of practical interest are the effects of the gas velocity, the geometry of the gas distributor, and the filling height.

The effects of these factors were observed by numerous investigations [27] or [44, 45], with various gas distributors uniformly arranged at the base of the bubble columns. Some results are shown as examples in Figs. 12 and 13.

As expected, the gas velocity has a dominating influence on floc destruction, since it increases the average energy input. For a constant superficial velocity, the velocity in the sparger, v_L, is found to have a distinct influence (see right-hand diagram in Fig. 13). However, this influence is weak at high gas velocities, and is mainly present only within a certain range of v_L. The effect of outlet velocity of the gas even at relatively low dynamic pressures $\varrho_G v_L^2/2$ can be explained by the assumption of high energy densities in free jets corresponding to Eq. (15).

An increase in the hole diameter d_L with otherwise constant velocities v_L in the gas distributor leads initially to a decrease in the stress within the d_L range from 0.2 mm to 0.5 mm, and then again to an increase over the range from

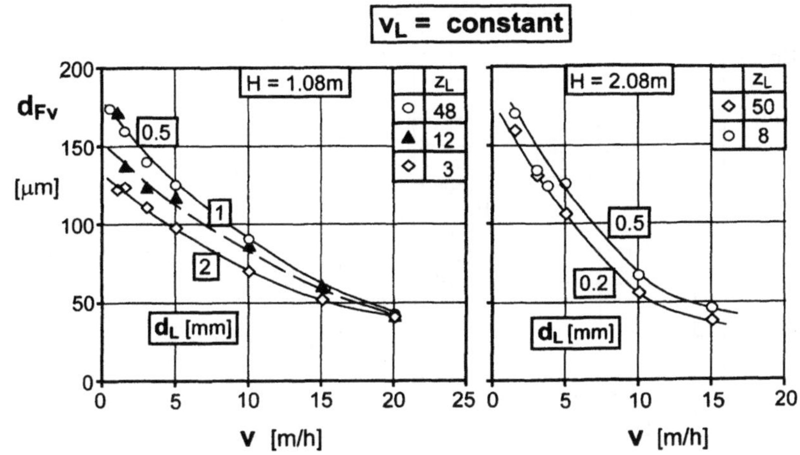

Fig. 12. Influence of superficial gas velocity v and hole diameter d_L of sparger on reference floc diameter d_{Fv} at constant hole velocities v_L for bubble columns; D = 0.4 m

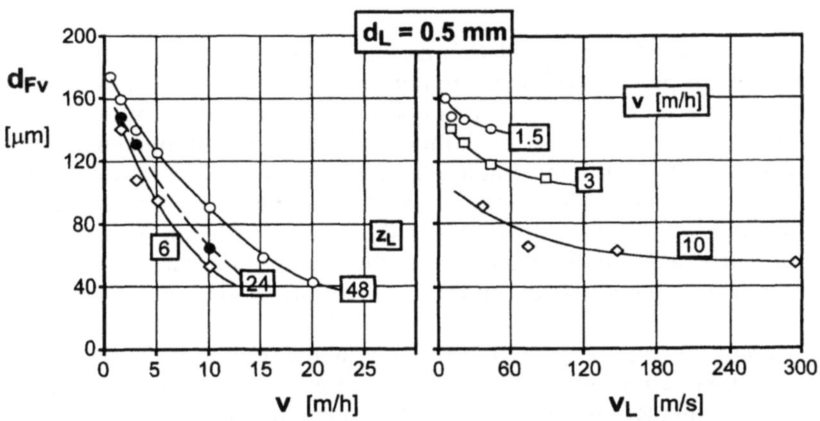

Fig. 13. Influence of superficial gas velocity v (left hand diagram), hole velocity v_L (right hand diagram) and hole number z_L of sparger on reference floc diameter d_{Fv} for bubble columns; D = 0.4 m; H = 1.08 m

0.5 mm to 2 mm (see Fig. 12). This indicates that the bubble size (see [27]) evidently plays a part. The larger bubbles with $d_B > 1-2$ mm, because of their increasingly more pronounced ellipsoidal shape with increasing bubble size, effect an increasing wobbling motion and this motion leads possibly to intensified bubble-floc contact. For $d_L < 0.5$ mm, on the other hand, it may be assumed that the destructive effect of the bursting of the bubbles at the liquid surface becomes increasingly important, which according to theoretical investigations [34] intensifies as the bubble size decreases.

The effects of bubble formation and of the bursting of bubbles at the surface become clear if filling height H is varied at high gas velocities v (see Fig. 14). At

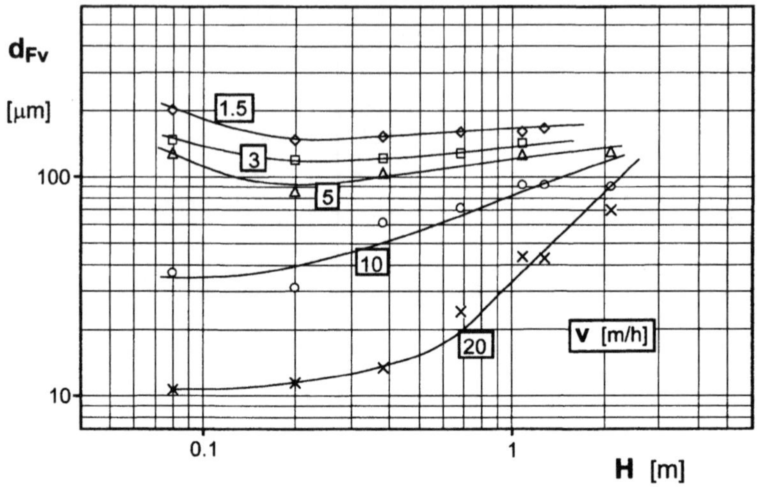

Fig. 14. Influence of filling height H on reference floc diameter d_{Fv} for bubble columns; $D = 0.4$ m; $d_L = 0.5$ mm; $z_L = 48$

$v = 20$ m/h for $H > 0.5$ m the relation $d_{Fv} \sim H$ is found, which suggests that the gas throughput per unit volume is important because $q/V \sim v/H$. This proportionality exists if the number of bubbles generated and escaping per unit volume determines the stress, i.e. if independently of the filling height and the operating conditions the maximum possible contact of the particles with the phase interface occurs. This condition is evidently no longer satisfied at lower gas velocities $v < 20$ m/h and with lower fill heights $H < 0.5$ m. This can be explained if it is assumed that the particle/bubble contact that determines the disintegration process is weak under these conditions because of less mixing motion with $H = 0.5\ \mathrm{m} < D$ and lower gas velocities.

In view of the importance of the particle/bubble contact, it may be assumed that the stress acting on the particles during gas sparging is determined by electrostatic interactions as well as by hydrophobic and hydrophilic interactions, which are determined by the nature of the liquid/solid system. The use of Pluronic as additive leads to the reduction of destruction process [44, 47] possibly due to less bubble/floc contact which is also described by Meier et. al. [67].

A definite correlation of the results of the measurements can be achieved by using the adiabatic compressor power per unit volume of reactor according to Eq. (10) which is shown in Fig. 15 [27]. The experimentally determined loss factor ζ is required in Eq. (10). The measured data for spargers with holes $d_L = 0.2 - 2$ mm can be correlated with Eq. (24).

$$\zeta = 7000/\mathrm{Re}_G \quad \text{for} \quad \mathrm{Re}_G = v_L d_L / n_G \leq 3000 \qquad (24)$$

$$\zeta = 8.3/\mathrm{Re}_G^{0.15} \quad \text{for} \quad \mathrm{Re}_G \geq 3000$$

The ζ values are clearly greater than 1, which indicates a contraction of the gas jets due to the sharp-edged holes.

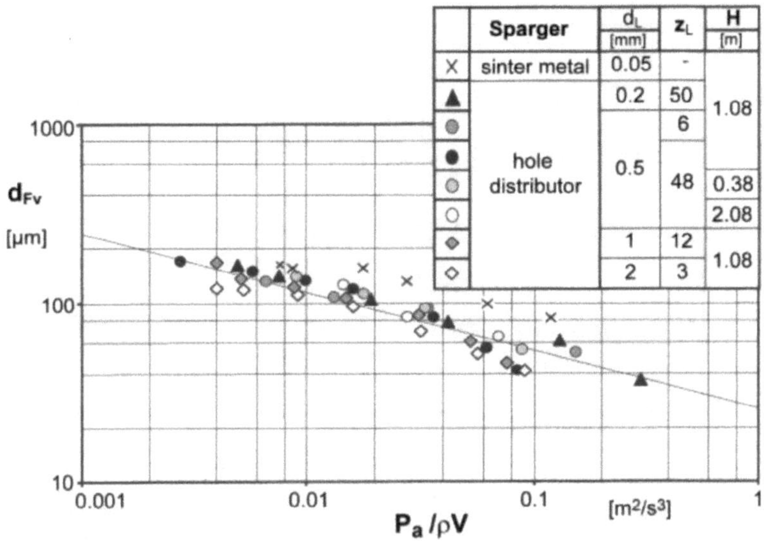

Fig. 15. Reference floc diameter d_{Fv} for bubble columns versus mass related adiabatic compressor power $P_a/\varrho V$

6.3.5
Loop Reactors

A number of investigations were carried out with loop reactors with different geometries, see [45], [27]. Two different arrangements of the gas distributor (gas introduced inside or outside the draught tube) and bottom distances of draught tube and the base were used. Two of the variants were selected according to the data of Blenke [69] to give the smallest flow losses. The results have shown the influence of the bottom spacing. Lower installation heights produce increased stress on the particles as a result of greater deflection losses. The increase in these losses was confirmed by parallel measurements, which showed a 25–50% decrease in the circulation time which means circulation velocity. This agrees with the simulations of Mechuk and Berzin [70] to determine the dissipation of energy in loop reactors. Even with the most favourable geometry, the stresses found in loop reactors were no smaller than those in bubble columns, which agrees with results reported by Hülscher [68] who made this observation during the cultivation of animal cells.

6.3.6
Comparison of Various Reactors

To compare the various reactor systems, the reference floc diameter was plotted as a function of stress (Fig. 11) and of energy density (Figs. 16 and 17).

While in Fig. 16 selected results are plotted against the average energy density $\bar{\varepsilon} = P/V\varrho$, in Fig. 17 all of the essential results for stirred tanks and bubble

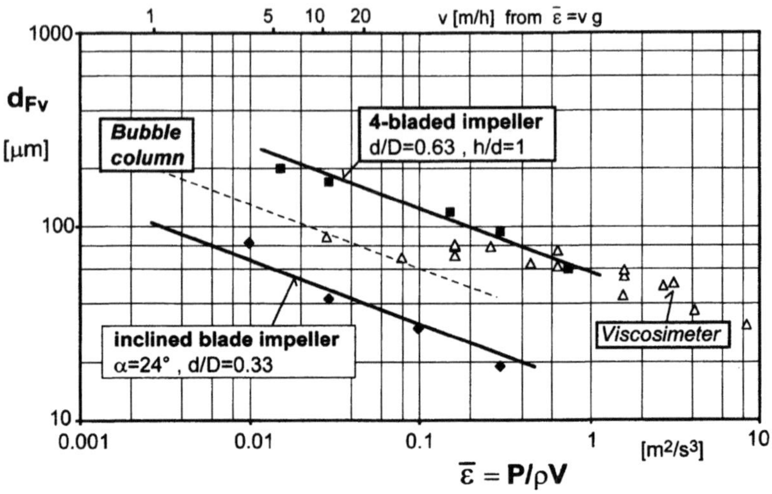

Fig. 16. Comparison of floc destruction in stirred vessels with baffles, bubble columns and viscosimeters

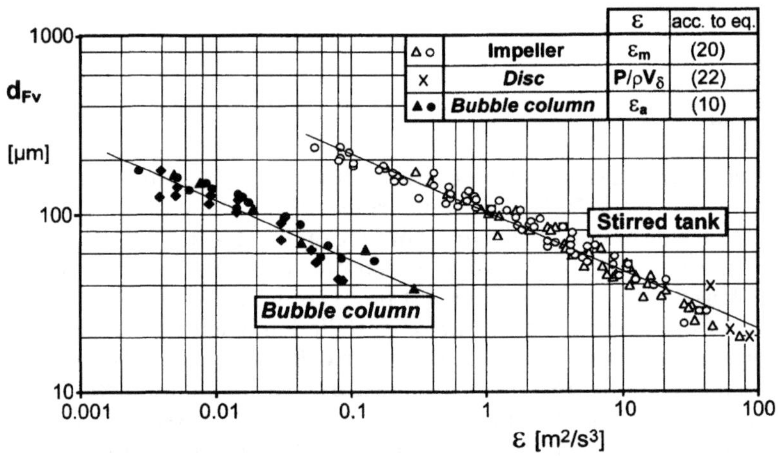

Fig. 17. Comparison of floc destruction stirred vessels with baffles and bubble columns

columns are shown. In Fig. 17 (also in Fig. 11) for the stirring reactors, the maximum energy dissipation ε_m according to Eq. (20) was used as variable.

It can be seen that for the same average power input, greater stresses are produced by gas sparging than by many impellers, Fig. 17. According to the comparison in Fig. 17, evidently zones exist in bubble columns in which the energy densities are 20 times higher than in a stirred tank. But the comparison on the basis of average power input in Fig. 16 shows that also impeller (for example small inclined blade impellers) exist which produce more shear than bubble columns.

Viscosimeters at higher average power input only cause stresses of the same order as that produced by low-shear impellers (Fig. 16).

6.4 Other Particle Systems

6.4.1 *Shake Flasks*

Ueda et al. [71] and Büchs et al. [72] used the dispersion behaviour in a two-phase mixture of toluene-carbon tetrachloride/water to evaluate the suitability of shake flasks. The systematic tests with the coalescence-inhibiting addition of lauryl ethylene oxide in [72], in shake flasks of various size ($V_t = 100 - 1000$ ml) with and without baffles, and in a stirred tank, permit conclusions on energy distribution in shake flasks. Comparison of the results from shake flasks and a stirred vessel shows that the energy distribution in shake flasks is almost uniform and, therefore, that the conditions are completely different from those in stirred fermenters. When, for example, standard disc stirrers with $d/D = 0.33$ are used, according to Eq. (20) the maximum energy dissipation near the stirrer is approx. 36 times greater than the average: $\varepsilon_m/\bar{\varepsilon} \approx 36$. As under normal operating conditions the mean volume-specific power input of shake flasks is in the same range as that of stirred fermenters [17, 72], biological cultures in shake flasks are subjected to substantially less stress than in stirred reactors. In addition, particularly in smaller sized shake flasks $V_t \leq 500$ ml, even with viscosities similar to that of water, turbulence is not fully developed. It is therefore to be expected that the transfer of shear-sensitive organisms from shake flasks to stirred fermenters will cause problems, and this has been confirmed by extensive practical experience.

6.4.2 *Baffled Stirred Tank*

To check the results obtained with the floccular system, comparative investigations were also carried out with the oil/water emulsion and the enzyme resin particle system [27] or [45, 47, 49]. The reactor and impeller geometry's were identical for these investigations.

The diagram in Fig. 18 shows direct comparisons with the corresponding results for the floccular system. The particle diameters d_{Fv} and d_E and the relative enzyme activity a/a_0 in Fig. 18 show similar patterns of variation as with the specific impeller power P/V. It is therefore appropriate to represent these results by means of the correlation function obtained for the floccular system according to Eq. (20). As in Fig. 9, a clear correlation of the results is found for both systems (see Figs. 19 and 20). It is thus clear that particle disintegration in a stirred tank with baffles follows a similar pattern for other particle systems.

The relationship d_{Fv}; $d_E \sim \varepsilon_m^{-b}$ found for the floccular system and the oil/water emulsion, with $b = -1/3$, confirms the theoretically derived Eq. (16b) where particle disintegration is determined by the turbulent eddies in the dissipation range.

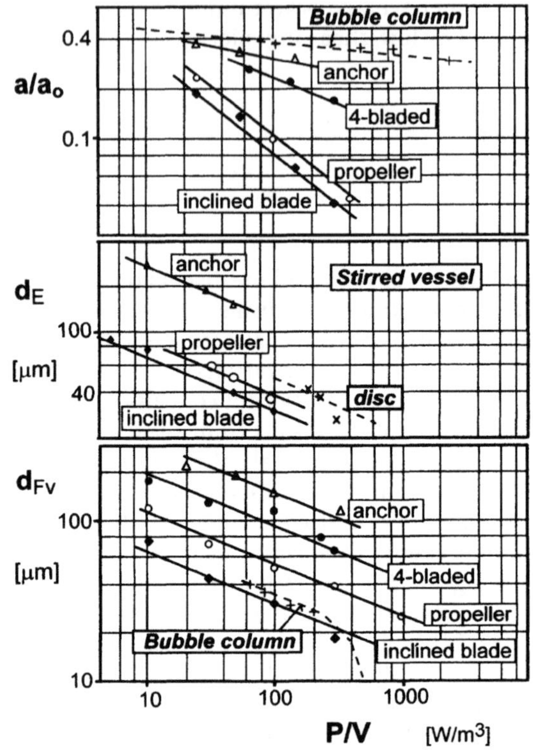

Fig. 18. Comparison of results from various particle systems for stirred vessel with baffles and bubble columns: Activity a/a_0 of Acylase resin after $t = 300$ h, equilibrium drop diameter d_E of silicon oil-water-surfactant emulsion and reference floc diameter d_{Fv} of floc system in dependency on specific power P/V; $H/D = 1$; $D = 0.15$ m; 0.4 m

6.4.3
Stirring with a Smooth Disc

In the case of stirring with a smooth disc, particle stresses found in the oil/water emulsion are similar and those in the enzyme resin lower than those in the floccular system (see Figs. 18, 19 and 20). This indicates that the enzyme particles are not subjected to the high energy density in the boundary layer of the disc to the same degree as in the floccular and emulsion system.

6.4.4
Bubble Column

With the oil/water/surfactant droplet system which was used, no investigations could be performed because of strong foaming. However, studies with water/kerosene emulsions are known from the literature. The results of Yoshida

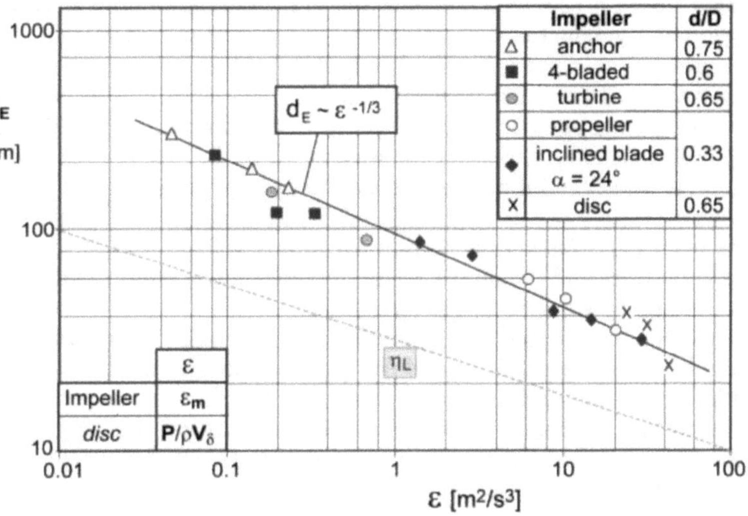

Fig. 19. Correlation of equilibrium drop diameter d_E of silicon oil-water-surfactant emulsion for stirred vessels; 4 baffles: w/D = 0.1; H/D = 1; D = 0.4 m

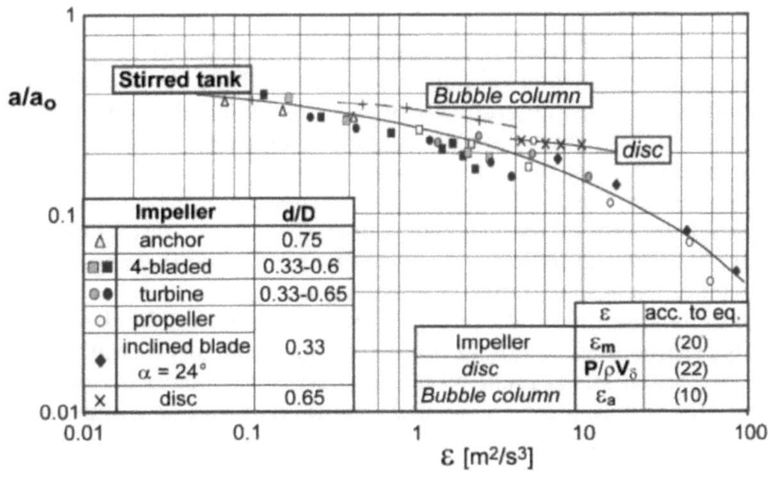

Fig. 20. Correlation of the residual activity a/a_o of the Acylase resin after t = 300 h stirring or sparging; stirred vessel: 4 baffles; w/D = 0.1; H/D = 1; D = 0.15 m; bubble column: hole sparger; d_L = 1 mm; z_L = 8; H/D = 1; D = 0.15 m

and Yamada [73] show that in the case of gassing, the same droplet size is obtained with only 30% of the average power input $\bar{\varepsilon} = P/\varrho V$ of a turbine agitator (d/D = 0.5; z = 12; 4 baffles). This is similar to our observation with floccular systems (see Fig. 17). If we assume that in [73] a standard turbine impeller with h/d = 0.2 was used the maximum energy dissipation ratio of $\varepsilon_m/\bar{\varepsilon} \approx 10.5$ can be calculated from Eq. (20) which means that in the bubble column approximately 35 higher energy densities takes place tan in the used stirred vessel.

In contrast to this, the enzyme resin is stressed less by gas sparging than by stirring (see Fig. 18 and 20). The same activity losses were observed first with 1 to 8 times greater specific adiabatic compression power $P_a/\varrho V$ than the maximum power density ε_m necessary for stirring. As in the case of the smooth disc, the effects of power input are only weak. The type of gas sparger and therefore the gas exit velocity are of no recognisable importance. The behaviour of the enzyme resin particles is thus completely different from that of the clay mineral/polymer flocs and the oil/water/surfactant droplet system, which are particularly intensively stressed by gas sparging.

This result makes it clear that particle stress is strongly dependent on the interaction between the particles and the interface, so that electrostatic and also hydrophobic and hydrophilic interactions with the phase boundary are particularly important. This means that the stress caused by gas sparging and also by boundary-layer flows, as opposed to reactors with free turbulent flow (reactors with impellers and baffles), may depend on the particle system and therefore applicability to other material systems is limited.

7 Laws of Scale for Stirred Tanks

To clarify the laws of scale, investigations were conducted in geometrically strictly similar tanks having volumes V = 20, 50, and 730 l. For the same specific power, both the floc disintegration kinetics (Fig. 3) and the reference floc diameter (see Fig. 21) produce similar numerical values for all three scales. The same was also found for the other material systems.

For comparison, Fig. 21 also shows the dependence of the results on the circumferential velocity of the impeller. Contrary to many assumptions in the literature, this diagram indicates that velocity is not a significant process parameter as far as the disintegration process is concerned.

At the same time, the results in Fig. 3 – because of the validity here of the relationships represented by Eqs. (25) and (26) – rule out any possibility that the number of circulations z_F made by the particles or the power per unit circulation flow P/q_F is important. In the first case there would be a dependence on scale according to Eq. (25), and in the second case there would also be a dependence on the circumferential velocity $u \sim nd$ according to Eq. (26).

$$z_F = \frac{t}{t_F} \propto \frac{q_F}{V} \propto \left[\frac{c_F}{Ne^{1/3}}\left(\frac{h}{D}\right)^{4/3}\right]\frac{(P/V)^{1/3}}{V^{2/9}} \tag{25}$$

$$\frac{P}{q_F} \propto \frac{Ne}{c_F}(nd)^2 \propto \left[\frac{Ne^{1/3}}{c_F(d/D)^{4/3}}\right]\left(\frac{P}{V}\right)^{2/3} V^{2/9} \tag{26}$$

The fact that the kinetics at the same power per volume are also independent of the reactor size, i.e. of the number of circuits made by the particles, points to

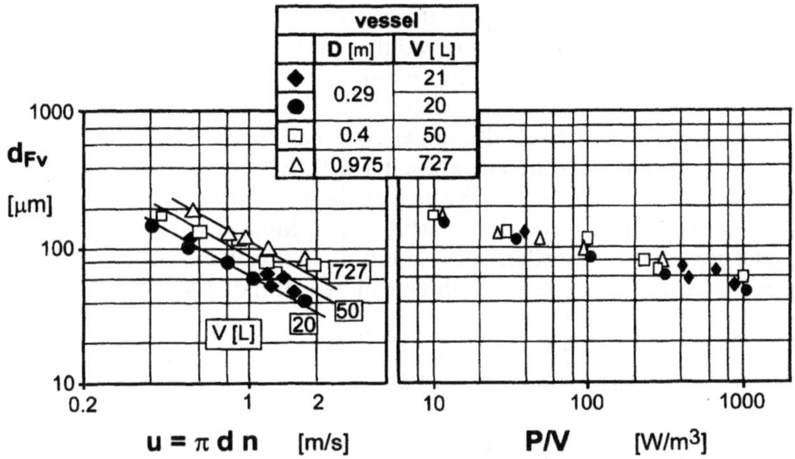

Fig. 21. Results of scale up investigations: Reference floc diameter d_{Fv} versus tip speed velocity u of impeller or the specific impeller power P/V; 4-bladed impeller; d/D = 0.33; h/d =; 4 baffles; w/D = 0.1; H/D = 1

the conclusion that the determining influence on disintegration is the total residence time in the zones of maximum stress,

$$t_{D\,tot} = z_F t_D = t\,\frac{V_D}{V} \tag{27}$$

which is independent of the circulation time $t_F = V/q_F$.

8 Particles Stress Equations Derived from Empirical Results

It could be shown (see Sect. 6) that in stirred vessels with baffles and under the condition of fully developed turbulence, particle stress can be described by Eqs. (2) and (4) alone. The turbulent eddys in the dissipation range are decisive for the model particle systems used here and many biological particle systems (see Fig. 2), so that the following equation applies to effective stress:

Stirred vessel with baffles:

$$\tau_t \approx 0.0676\,\rho d_P^2 \left(\frac{\varepsilon_m}{\bar{\varepsilon}}\right)\frac{\bar{\varepsilon}}{\nu} \text{ with } \left(\frac{\varepsilon_m}{\bar{\varepsilon}}\right) \text{ from equation (20).} \tag{28}$$

This equation should generally valid for all particle systems and working conditions with $(\rho_P - \rho)/\rho \ll 1$, $d_P/\eta_L < 6$ and $\Lambda/\eta_L > 125\ldots250$. The last condition of fully developed turbulent flow is very important. To small values Λ/η_L which mostly corresponds to Reynolds numbers $Re < 10^4$ (small reactors, higher viscosity's of media and small power input) leads to an distinct reduction of stress. That was observed by the investigations in [66] which were carried out with the aim to develop a small stirred apparatuses ($V \approx 1.5$ L, diameter $D = 0.1$ m) as a shear tester for animal cells. It was found that in the case of working conditions of $\Lambda/\eta_L \approx 120$ ($Re = nd^2/\nu \approx 8000$) approximately 25% less stress is exposed to particles than in a geometrical similar, technical reactor ($V = 80$ L, $D = 0.4$ m).

In the case of stirred vessels the values Λ/η_L can be calculated by the following equation using the geometry parameter d/D, H/D, the Newton number Ne, the Reynolds number $Re = nd^2/\nu$, the energy dissipation ratio $\varepsilon/\bar{\varepsilon}$ and the related macro scale Λ/d. For standard turbines e.g. Möckel [24] found the value $\Lambda/d \approx 0.08$ close to the impeller. Corresponding to this the maximum of the dissipation ratio $\varepsilon_m/\bar{\varepsilon}$ has to be used which can be estimated by Eq. (20).

$$\frac{\Lambda}{\eta_L} = \left[\frac{4}{\pi}\frac{Ne}{H/D}\left(\frac{d}{D}\right)^3\right]^{1/4} Re^{3/4}\left(\frac{\varepsilon}{\bar{\varepsilon}}\right)^{1/4}\left(\frac{\Lambda}{d}\right) > 125\ldots250$$

The stress caused by gas sparging and also by boundary-layer flows, as opposed to reactors with free turbulent flow (reactors with impellers and baffles), may depend on the particle system.

Much higher shear forces than in stirred vessels can arise if the particles move into the gas-liquid boundary layer. For the roughly estimation of stress in bubble columns the Eq. (29) with the compression power, Eq. (10), can be used. The constant G is dependent on the particle system. The comparison of results of bubble columns with those from stirred vessel leads to $G \approx \gg 1.35$ for the floccular particle systems (see Sect. 6.3.6, Fig. 17) and for a water/kerosene emulsion (see Yoshida and Yamada [73]) to $G \approx 2.3$. The value for the floc system was found mainly for hole gas distributors with hole diameters of $d_L = 0.2-2$ mm, opening area $A_L/A = z_L(d_L/D)^2 = (0.9\ldots80)10^{-5}$ and filled heights of $H = 0.4-2.1$ m (see Fig. 15).

$$\text{Bubble columns:}\quad \tau_t \approx G\varrho d_p^2 \frac{\bar{\varepsilon}_a}{\nu} \quad \text{with } \bar{\varepsilon}_a \text{ from equation (10).} \tag{29}$$

Viscosimeter flow produces less stress than technical reactors (see Sect. 6.3.3). From the results with the floccular particle system it can be derived the following relationship (30). It estimates the turbulent stress τ_t of a technical, fully baffled stirred reactor which leads to the same damage of particles as the viscosimeter flow with the shear stress τ.

$$\tau_t \approx 0.68\ [N^{1/2}/m]\ \sqrt{\tau} \quad \text{for } t > 1\ N/m^2 \tag{30}$$

9 Stress in Biological Particle Systems

A few exemplary results obtained with biological systems are discussed in the following and compared with the above-mentioned basic results of model particle systems.

Figure 22 shows a comparison of results from model particle systems and literature data with biological systems in stirred vessels. The dependency of particle diameter on maximum energy dissipation: $d_P \sim \varepsilon_m^{-1/3}$ of yeast and BHK

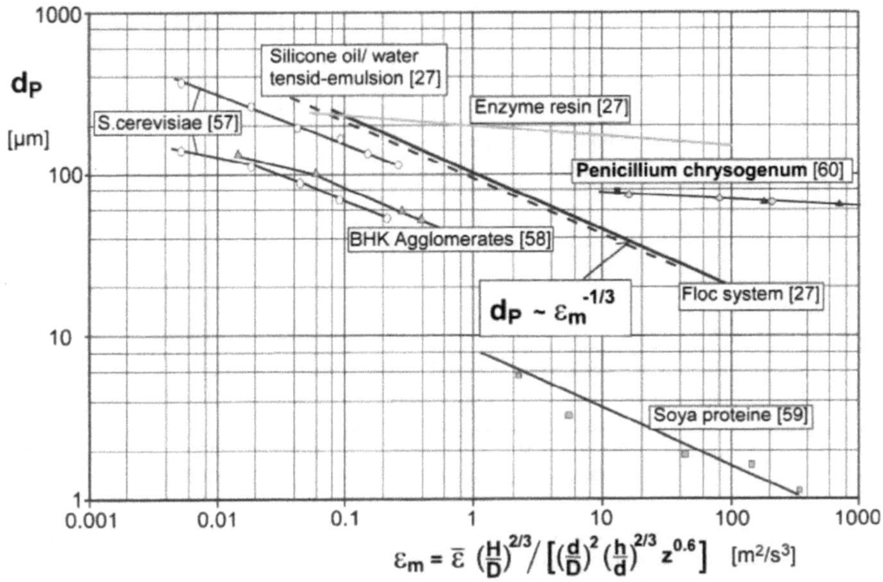

Fig. 22. Dependency of average particle diameter d_P on maximum energy dissipation ε_m of impeller systems with baffles by stirring of biological and model particle systems; explanations see Tables 3 and 4

agglomerates and also of soya protein aggregates, is similar to that of the floccular and emulsion model systems. Equation (20) was used here for the calculation of maximum energy dissipation ε_m.

It confirms again the theoretically derived Eq. (16b) and means that also particles which are much smaller as the smallest turbulent eddies ($\approx 3\ \eta_L$, see Fig. 2) are disrupted by the turbulent eddies of the dissipation range. For the calculation of stress has to be used the Reynold's stress Eq. (2) and not Newton's law (1).

9.1 Microorganisms

A number of tests have been carried out on the stress on mycelium-forming microorganisms, whereby the works of Jüsten [60], which will be dealt with in more detail in the following, are particularly extensive.

With the knowledge of the basic tests on particle stress in model particle systems [45] and [47], Jüsten [60] selected similar stirrer types and operating conditions for his tests on stress on the mycelial microorganism, Penicillium chrysogenum. These experiments were carried out with sufficient oxygen supply so that the results may only be interpreted as due to different stress.

The square root of projected area of organism d_P, the growth rate μ and productivity P/Xt show an influence of power input and impeller geometry (see left-hand diagram in Figs. 23 and 24). These results can be correlated (see right-

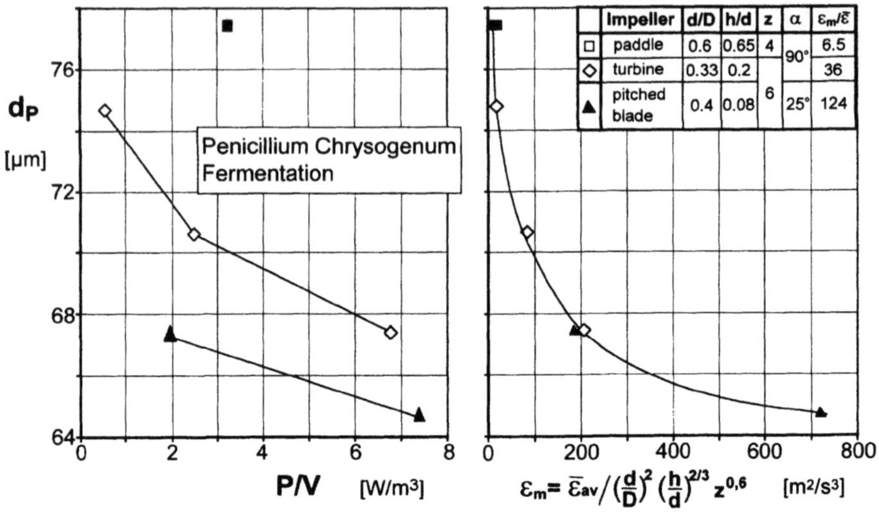

Fig. 23. Average particle size d_P after t = 120 h stirring for various impeller types and working conditions (left hand diagram: data from [60]) and correlation with the maximum energy dissipation ε_m (right hand diagram); stirred bioreactor with 4 baffles; V = 6 L; D = 0.2 m; H/D = 0.96; $z_I = 1$

hand diagram in Figs. 23 and 24) by using the maximum energy dissipation, Eq. (20) and the shear stress Eq. (28). For $\bar{\varepsilon}$ and τ_t the time average quantities $\bar{\varepsilon}_{av}$ and τ_{tav} over the relevant time frame t = 0 to t = T of fermentation were taken:

$$\bar{\varepsilon}_{av} = \frac{1}{T}\int_0^T \frac{P}{\varrho V}\,dt \quad \text{or} \quad \tau_{av} = \frac{1}{T}\int_0^T \tau_t\,dt \tag{31}$$

For the calculation of shear stress, the time-dependent impeller power, particle diameter d_P and viscosity ν according to $\nu = K/\varrho \cdot \gamma^{m-1}$ with the representative shear gradient $\gamma = \sqrt{\varepsilon_m/\nu}$ for the non Newtonian broth (see equation (17) [28]) were used.

With the increase of shear stress, which is mostly changed by energy dissipation, the projected area decreases slightly. This fragmentation leads to a stimulation of growth but unfortunately also to a decrease in the overall cell mass X related productivity P/X t.

The possibility of correlating these fermentation parameters with the turbulent stress equation shows again that obviously similar relationships exist for both the biological systems and the model particle systems used here.

Jüsten [60] took the quantity $P/(V_I\, t_F) \sim z_F\, P/V_I$ for the correlation of his results which means that in larger reactors less stress takes place at the same power input P/V_I. The assumption that the number of circulations through the impeller zone z_F is decisive for the load of organisms was put forward by Smith, Lilly, Fox [74]. This parameter cannot be important since the circulation frequency $1/t_F \sim z_F$ is normally small in comparison to the frequency of turbulent fluctuations f_t. That means that the dynamic stress arising from turbulence not only occurs once per circulation. According to Eq. (27) it is acting over the total residence time of organisms in the high-shear region.

The turbulent fluctuation frequency can be estimated by means of turbulent measurements. Möckel [24] found that the wave number $k = 2\Pi f_t / u$ in the interesting dissipation range is $k > k_o$ with the limiting value $k_o = (0.1 \ldots 0.2)\eta_L$. From this becomes the frequency to $f_t \geq (0.016 \ldots 0.032) u/\eta_L$. An important measure should be the related number of turbulent fluctuation z_t/z_F which occur during the residence time of particles $t_I = V_I/q_F$ inside the fictive impeller volume V_I at one circulation. It follows to:

$$\frac{z_t}{z_F} \geq (0.004 \ldots 0.008) \frac{d}{\eta_L} \qquad (32)$$

The ratio z_t/z_F is in technical reactors much higher than 1. It becomes, e.g. also for a small scale reactor of $V \approx 100L$ ($H/D = 2{:}D = 0.4$ m) equipped with three turbines ($d/D = 0.3$) and working at a average impeller power per mass of only $\bar{\varepsilon} = 1 m^2/s^3$ in media with water like viscosity to $z_t/z_F \geq 36 \ldots 72$. The maximal energy dissipation in the impeller zones, required for the calculation of length scale of turbulence η_L, was here taken from Eq. (20).

Other investigations of Jüsten [60] in regard to sparging with a superficial velocity of $v = 30$ m/h ($\bar{\varepsilon}_a = P_a/\varrho V \approx 0.082 m^2/s^3$) show no effects on morphology. Although these investigations, which were performed in a very small bubble column ($D = 0.07$ m; $H = 0.21$ m), were not carried out under technically

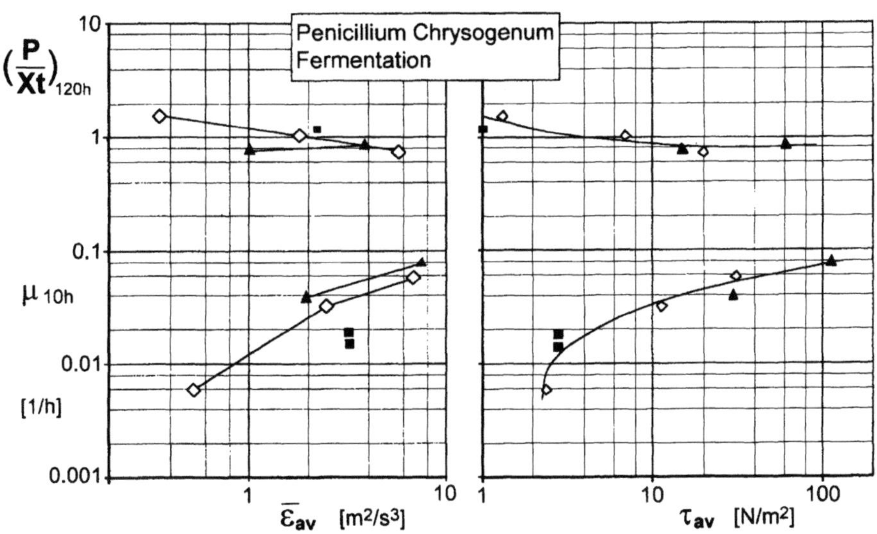

Fig. 24. Influence of impeller type and working conditions on productivity P/Xt and growth rate μ of Penicillin G batch fermentations with Penicillium chrysogenum (left hand diagram: data from [60]) and correlation by using the turbulent stress τ_t corresponding to Eq. (28) (right hand diagram); symbol explanations see Fig. 23

relevant conditions, they indicate that the mycelia had no strong interactions with the gas-liquid phase boundary. That is in agreement with the experience in large scale citric acid fermentations of Aspergillus niger, which is often performed in bubble columns. Aspergillus niger reaches maximum productivity if it grows as a pellet. This pellet formation is not hindered at very high superficial velocities of $v = 150$ m/h which corresponds to a relatively high dissipation rate of $\bar{\varepsilon}_a \approx 0.4\ m^2/s^3$.

9.2 Animal Cells

Animal cells are more sensitive to shearing than micro-organisms because they have no cell wall. Therefore in many shearing investigations used animal cells. But unfortunately most of these investigations were carried out in laminar flow systems like viscosimeters and channels. As mentioned above (see Sect. 4.2), the relationships are not comparable to technical reactors such as sparged impeller vessels.

The experimental results for hybridoma and protozoa cells given as examples in Fig. 25 indicate that much higher stress (4 to 30 times) is required under laminar flow conditions of viscosimeters than in stirred vessels to achieve the same death rate k_d. Here the death rate k_d is defined as first order deactivation constant: $k_d = 1/t \cdot \ln(N_o/N)$, where N_o is the initial and N the time-dependent number of living cells in special deactivation experiments under otherwise optimal living conditions. The stress in Fig. 25 was calculated with Eq. (28) for stirred vessels and with Eq. (1) for the viscosimeter. Our own results for hybri-

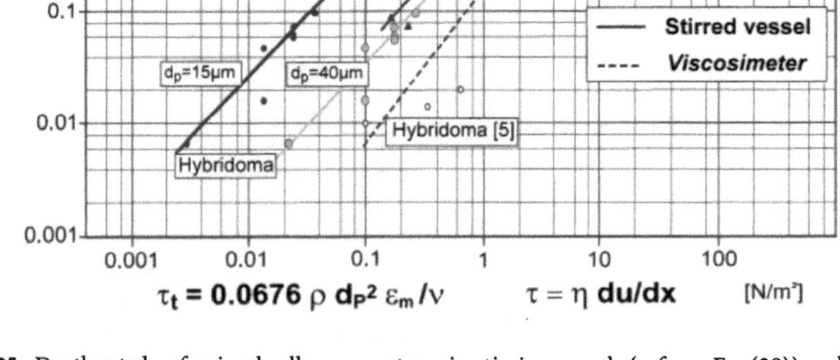

Fig. 25. Death rate k_d of animal cells versus stress in stirring vessels (τ_t from Eq. (28)) and viscosimeters (τ from Eq. (13))

doma cells and stirred vessels are presented with the average agglomerate diameter of $d_P \approx 40\ \mu m$ and the average diameter of single cells $d_P \approx 15\ \mu m$ since both aggregates and single cells were present in the cell culture which was used. If the agglomerates withstand the shear stress, the cells in aggregates are exposed to higher stress than single cells, due to the larger particle diameter (see Eq. (28)).

The results in Fig. 25 for hybridoma cells show that due to the low growth rate, which lies in the range of $\mu \leq 0.02/h$, animal cells could only be cultivated under very moderate stress conditions of $\tau_t < 0.005 - 0.05\ N/m^2$. That means that, also for very low-shear impellers such as large-blade impellers, the average power input has to be limited to $P/V \leq 30 - 50\ W/m^3$.

On the other hand animal cells are especially sensitive as regards sparging. Obviously the cells are adsorbed at the gas liquid boundary layer and subjected to the most stress in the region of bubble formation at the sparger and bubble bursting at the liquid surface.

Therefore an influence of filled height and bubble size and consequently of sparger design (hole diameter and number) can be assumed (see Sect. 6.3.4). It is important to know under which sparging conditions the lowest stress can be expected. For the floccular model system described above e.g., both mass transfer and shearing measurements were carried out. Figure 26 gives these data, which allow a preliminary estimation of this relationship. The highest k_La values and the highest particle diameter (corresponding to the lowest stress) were obtained with fine bubble aeration realised by a sparger with a small hole or pore diameter.

It is to assume that similar relationships are valid for sparging of animal cell cultures.

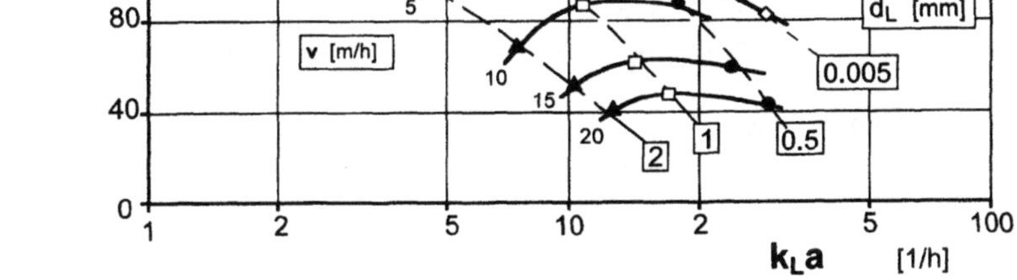

Fig. 26. Reference floc diameter d_{Fv} of floc particle system versus mass transfer coefficient k_La for bubble columns with different gas spargers: $H = 1.08$ m; $D = 0.4$ m

9.3 Enzymatic Reaction with Immobilised Enzymes

For bio-transformation processes, immobilised enzymes are often used because their activity persists over a longer period of time than that of free enzymes. The reduction of enzyme activity in enzymatic reactors is a consequence of energy dissipation by sparging and stirring, which is required for instance for oxygen transport or realisation of constant reaction conditions as regards temperature and pH. In the other hand low and high pH-values leads also to a decrease of enzyme activity and increase the stress sensitivity.

As an example the deactivation of immobilised Pen G acylase, which catalyses the reaction of Pen G to 6-Aminopenicillanic acid and Phenylacetic acid, was studied. This enzyme was covalently bound on an ion-exchanger and cross-linked by glutaric aldehyde. To maintain a high reaction velocity, a neutral pH value (removal of Phenylacetic acid) and therefore the supply of NaOH and stirring for distribution of the base are required.

The experimental results in Fig. 27 show the influence of the reactor system (see Fig. 28) on the disintegration of enzyme activity. It was found that the low-stress bladed impeller results in less activity loss than the propeller stirrer which causes much higher maximum energy dissipation ε_m. The gentle motion the blade impeller produces means that stress is so low that its disadvantage of worse micro mixing in NaOH (in comparison with the propeller) is more than compensated.

Short fixed beds are better than stirred reactors. Here a bed with a height of 10 cm arranged in a circulation loop for the control of pH was used. Although the average dissipation rate in the fixed bed with $\bar{\varepsilon} = 7.2$ and 64 m^2/s^3 is much

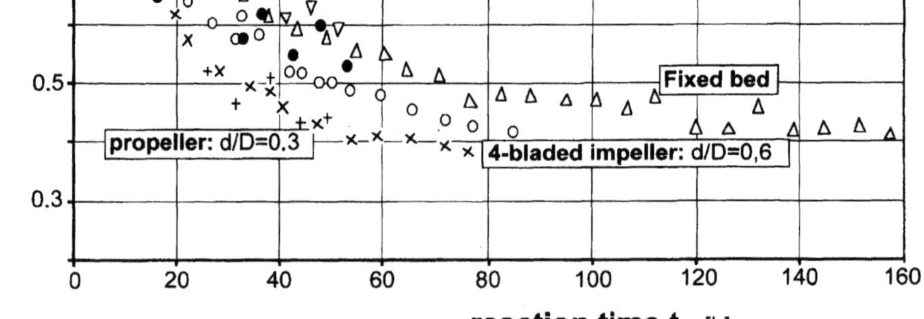

Fig. 27. Activity loss a/a_o of Acylase enzym resign with the reaction time t of the enzymatic deaccylation of Penicillin G to 6-Aminopenicillanic acid (reactor design see Fig. 28)

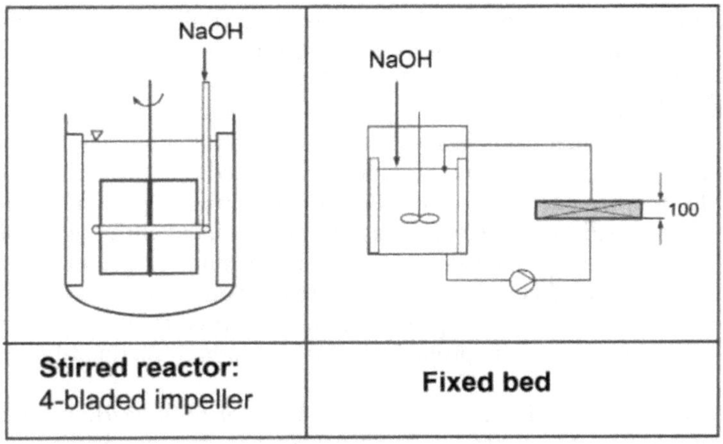

Fig. 28. Reactor design for the enzymatic deaccylation of Penicillin G to 6-Amino-penicillanic acid

higher than even the maximum dissipation in the stirred vessel with $\bar{\varepsilon} = 0.125 \ldots 6{,}86\ m^2/s^3$, a smaller decrease of enzyme activity is observed. That may be explained again by the different flow conditions. In contrast to the stirred vessel, where there is turbulent flow, the flow in the fixed bed is purely laminar.

10 Conclusions

Many results with model systems and also biological particle systems indicate that the stress in technical bioreactors, in which turbulent flow conditions exist, could not be simulated by model studies in small bioreactors, where no fully turbulent flow exists, and especially with laminar flow devices such as viscosimeters, tubes or channels.

The comparison of the results obtained from model particle systems with experience of biological systems shows a similar tendency on many points. Therefore it proved to be very advantageous for the basic investigations, especially for the comparison of different reactor types, to use suitable model particle systems with similar properties to those of biological material systems. This permitted the performance of test series under technically relevant operating conditions, similar to those prevailing in bioreactors, in a relatively short time. The results are more reproducible than in biological systems and therefore permit faster and more exact optimization of reactors.

Following basic tests of this kind, as a rule only a punctual comparison with the biological material system in question is required. This procedure is recommended in this report.

The presented results for systematic studies on hydrodynamic stress in shake flasks, baffled stirred tanks, reactors in which boundary layer flow predominates (e.g. stirred tank with a smooth disc or unbaffled stirred tank), viscosimeters, bubble columns, and gas-driven loop reactors allow the selection of bioreactors according to the particle stress criterion.

For reactors with free turbulent flow without dominant boundary layer flows or gas/liquid interfaces (due to rising gas bubbles) such as stirred reactors with baffles, all used model particle systems and also many biological systems produce similar results, and it may therefore be assumed that these results are also applicable to other particle systems. For stirred tanks in particular, the stress produced by impellers of various types can be predicted with the aid of a geometrical function (Eq. (20)) derived from the results of the measurements. Impellers with a large blade area in relation to the tank dimensions produce less shear, because of their uniform power input, in contrast to small and especially axial-flow impellers, such as propellers, and all kinds of inclined-blade impellers.

In reactors with predominant boundary-layer or with gas/liquid interfaces due to bubble flow, the stress acting on particles is strongly dependent on the tendency of the particles to move into the boundary layers. Whereas flocs, oil/water/surfactant droplet systems and also animal cells can be much more strongly stressed in these reactors than in baffled stirred tanks, the opposite is found for the enzyme resin, and also for microorganisms such as Penicillium chrysogenum. It can therefore be concluded that particles with different surface properties (e.g. charge-dependent, hydrophobic, hydrophilic), which result in different interactions with the interfaces and phase boundaries, also suffer different particle stresses in boundary-layer flows. Thus the scope for application of the results to other particle system is limited in such reactors. In this case more special investigation with the original material system must be performed.

References

1. Liepe F (1988) Verfahrenstechnische Berechnungsmethoden. Part 4, VCH, Weinheim
2. Kolmogoroff A (1958) Collected works on the statistical theory of turbulence. Akademic Verlag, Berlin
3. Middler M, Finn K (1966) Biotech Bioeng 8:71
4. Tramper J, Willams JB, Joustra B, Vlak JM (1986) Enzym Microb Technol 8:33
5. Smith G, Greenfield F, Randerson D (1987) Biotechnol Techniques 1:39
6. Goldblum S, Bae Y, Hink WF, Calmers JJ (1990) Biotechol Prog 6:383
7. Dunlop EH, Namdev PK, Rosenberg MZ (1994) Chemical Engineering Science 49:2263
8. Augstein DC, Sinskey AJ, Wang DIC (1971) Biotechnol Bioeng 13:409
9. Reese ET, Ryu DY (1980) Enzym Microb Technol 2:239
10. Strathopulos NA, Hellums JD (1980) Biotechnol Bioeng 27:1021
11. Ludwig A, Kretzmer G (1992) Enzyme Microb Technol 14:209
12. Märkl H, Bronnemeier R (1982) Biotech Bioeng 24:553
13. Loughlin PFM, Malone DM, Murtagh JT, Kieran PM (1998) Biotech Bioeng 58:595
14. Sumino Y, Akiyama S, Fukuda H (1972) J Ferment Technol 50:203
15. Kato Y, Hiraoka S, Tada Y, Shirai S, Ue T, Koh ST, Yamaguchi T (1995) Kogaku-Ronbunshu 21(2):365
16. Kato Y, Hiraoka S, Tada Y, Koh ST, Lee YS (1996) Trans Ichem E 74:451
17. Büchs J, Maier U, Milbradt C, Toels B (1999) submitted Biotech and Bioeng
18. Henzler HJ, Schedel M (1991) Bioprocess Engineering 7:123

19. Zoels B (1992) Quantifizierung und Optimierung der Screening-Bedingungen im Schüttelkolben Master Thesis Fachhochschule Mannheim
20. Henzler HJ (1978) Homogenisieren von Flüssigkeiten und Gasen. VDI-Forschungsheft 587
21. Judat H (1976) Dissertation, Universität Berlin
22. Henzler HJ (1982) Chem Ing Tech 54:461
23. Henzler HJ (1998) Mischen und Rühren. GVC-Tagung Baden-Baden
24. Möckel HO (1978) Dissertation, Technische Hochschule, Köthen
25. Laufhütte HD(1986) Dissertation, TU München
26. Geissler RK (1991) Dissertation, TU München
27. Henzler HJ, Biedermann A (1996) Chem Ing Tech 68:1546
28. Henzler HJ, Kauling J (1993) Bioprocess Engineering 9:61
29. Aunins, Henzler HJ (1993) Aeration in Cell Culture Bioreactors. In: Rehm HJ, Reed G, Pühler A Stadler P Biotechnology VCH 3:219
30. Handa A, Emery AN, Spier RE (1987) Develop Biol Stand 66:241
31. Handa-Corrigan A, Emercy AN, Spier RE (1989) Enzyme MicrobTechnol 11:230
32. Newitt DM, Dombrowski N, Knelman FH (1954) Trans Instn Chem Engrs 32:244
33. Garner FH, Ellis SRM, Lacey JA (1954) Trans Instn Chem Engrs 32:244
34. Boulton-Stone JM, Blake JR (1993) J Fluid Mech 254:237
35. Briones MA, Brodkey RS, Chalmers JJ (1994) Chem Eng Sci 14:2301
36. Briones MA, Chalmers RS (1994) Biotech Bioeng 44:1089
37. Naue G, Liepe F, Maschek HJ, Reher, EO, Schenk R (1988) Technische Strömungslehre I, VEB Deutscher Verlag für Grundstoffindustrie, Leipzig
38. Reiter M (1983). Untersuchungen zur Flokkulation im Zylinderrühreaktor und im durchströmten Rohr. Dissertation, Universität Berlin
39. Taylor G (1936) Proc R Soc London A157:546
40. Baldyga J, Bourne JR, Zimmermann B (1994) Chem Eng Sci 49:1937
41. Baldyga J, Bourne JR, Gholap RV (1995) Chem Eng Sci 50:1877
42. Tralles S (1998) Entwicklung eines Testsystems zur Bestimmung der Scherbeanspruchung in Bioreaktoren. Master Thesis, Universität Paderborn
43. Hoffmann J, Tralles S, Hempel DC (1992) Chem Ing Tech 64:953
44. Biedermann A, Henzler HJ (1994) Chem Ing Tech 66:209
45. Biedermann A (1994) Scherbeanspruchung in Bioreaktoren. Dissertation Universität Köln
46. Bücher K (1993) Scherbeanspruchung in Bioreaktoren. Master Thesis Universität Paderborn
47. Jüsten P (1993) Scherbeanspruchung in Bioreaktoren. Master Thesis Technische Hochschule Aachen
48. Keßler M (1994) Scherbeanspruchung in Bioreaktoren. Master Thesis Universität Köln
49. Schumacher C (1994) Partikelbeanspruchung in Reaktoren. Master Thesis Technische Hochschule Aachen
50. Hoffmann J, Büscher K, Hempel DC (1995) Chem Ing Tech 67:210
51. Hoffmann J (1994) Scherkräfte in Bioreaktoren. Dissertation Universität Paderborn
52. Bücher K (1997) Bestimmung von mechanischen Beanspruchungen in Zweiphasensystemen. Dissertation Universität Paderborn
53. Ayazai Shamlou P, Makagiansar HY, Ison AP, Lilly MD (1994) Chem Eng Sci 16:2631
54. Paul GC, Thomas CR (1998) Advances in Biochem Eng Biotechnology 60:1
55. Pons MN, Vivier H (1998) Advances in Biochem Eng Biotechnology 60:61
56. Schügerl K, Gerlach SR, Siedenburg D (1998) Advances in Biochem Eng Biotechnology 60:195
57. van Hamersfeld EH, van der Lans RGJM, Luyben KCAM (1997) Biotech Bioeng 56:190
58. Moreira JL, Cruz PE, Santana CS, Aunins JG (1995) Chem Eng Sci 50:2747
59. Shamlou PA, Stravrinides N, Titchner-Höcker N, Hoore M (1994) Chem Eng Sci 49:2647
60. Jüsten P (1997) Dependence of Penicillium Chrysogenum growth, morphology, vacuolation and productivity on impeller type and agitation intensity. PHD Thesis University of Birmingham

61. Henzler HJ, Kauling J (1985) Scale up of mass transfer in highly viscous liquids. Preprints of 5th European Conference on Mixing, Würzburg
62. Dechema Study for research and development: "Standardization and equipment recommendations for bioreactors and peripheral devices" (1990)
63. GBF Text No. 4 (1987) Bioreactors, a guideline for the user. GBF Verlag Braunschweig
64. Zhou G, Kresta SM (1996) Chem Eng Sci 44:2207
65. Schlichting H (1968) Boundary-Layer Theory. McGraw-Hill New York
66. Flagmeyer A, Konstantinov K, Henzler HJ submitted in Biotechnol and Bioeng
67. Meier SJ, Hatton TA, Wang DIC (1999) Biotechnol and Bioeng 62:468
68. Hülscher M (1990) Fortschritts Berichte VDI Z, Series 3, no. 339, VDI, Düsseldorf
69. Blenke H (1979) Adv Biochem Eng 13:121
70. Merchuk JC, Berzin I (1995) Chem Eng Sci 50:2225
71. Ueda K, Takebe H, Takahashi J, Nomoto M (1971) Ferment Technol 49:981
72. Büchs J, Zoels B (1999) Evaluation of the Maximum Specific Power Consumption in Shaking Bioreactors, Proceedings of 3th International Symposium on Mixing in Industrial Processes (ISMIP-3) 19–22 Sept 1999 Japan
73. Yoshida F, Yamada T (1971) Fermen Technol 49:235
74. Smith JJ, Lilly, MD, Fox RI (1990) Biotech Bioeng 35:1011
75. Smith GC, Greenfield FP, Randerson DH (1987) Biotechnol Techniques 1:39
76. Abu-Reesh I, Kargi F (1989) Journal of Biotechnology 9:167
77. Cherry RS, Known KJ (1990) Biotechnol and Bioeng 36:563

Received September 1999

The Engineering Effects of Fluids Flow on Freely Suspended Biological Macro-Materials and Macromolecules

S.S. Yim · P.A. Shamlou

The Advanced Centre for Biochemical Engineering, Department of Biochemical Engineering, University College London, Torrington Place, London, WC1E 7JE, United Kingdom
E-mail: p.shamlou@ucl.ac.uk

The manufacture of many biotechnologically important products requires consideration of the physical breakage and biochemical degradation pathways at all stages during processing, storage and transportation. The engineering flow environment in most items of bioprocess equipment has long been recognised as a key factor in determining these pathways and is the focus of the present review. Because of its industrial significance, the detrimental effects of the engineering flow environment on freely suspended bioparticles have been the subject of many scientific investigations over the past few decades. There is a general consensus of opinion that fluid shear and elongational stresses are the two main breakage pathways of relevance to processing of most biomaterials. An additional degradation pathway has also been identified involving significant losses of biological activity of macromolecules at gas-liquid, gas-solid and liquid-liquid interfaces. In such cases, the engineering flow field is shown to have a secondary role in determining the kinetics of inactivation. An equally important consideration in the optimisation of the relevant unit operations is the biomechanical integrity of the flow sensitive material. The biomechanical and biorheological parameters that determine the integrity of biomaterials are poorly defined, their evaluations present future research challenges and are of immediate engineering significance.

Keywords. Shear effects, Biological macromaterials, Process equipment

Advances in Biochemical Engineering/
Biotechnology, Vol. 67
Managing Editor: Th. Scheper

Abbreviations

A	annular area between a pair rotating cylinders (m^2)
a	strain rate parameter (s^{-1})
B	breadth (m)
b	distant between neighbouring cells (m)
C	empirical constant (–)
C_{max}	maximum volumetric concentration ($kg\,m^{-3}$)
C_v	volumetric concentration ($kg\,m^{-3}$)
D	diameter (m)
De	Deborah number (t_e/t_p)
D_{imp}	impeller diameter (m)
D_{in}	inner rotating cylinder diameter (m)
D_{noz}	nozzle diameter (m)
d	spherical cell diameter (m)
d_{50}	precipitate diameter for 50% oversize (m)
d_{90}	precipitate diameter for 90% oversize (m)
E	Young's modulus of elasticity (Pa)
f	Fanning friction factor (–)
f_{app}	mean apparent friction factor (–)
f_p	theoretical function of viscosity ratio (–)
G	shear modulus of elasticity ($N\,m^{-2}$)
K	impeller constant (–)
k	relaxation time constant (s)
L	length (m)
L_p	pipe length (m)
M	deformation (–)
M_c	critical deformation (–)
N	rotational speed (rps)
P	impeller power input (W)
ΔP	pressure difference (Pa)
p	viscosity ratio (–)
Q	volumetric flow rate ($m^3\,s^{-1}$)
R	dimensionless radius (–)
Re	Reynolds number (–)
r	radius (m)
r_{in}	inner cylinder radius (m)
r_o	arbitrary starting point of the cell (–)
r_{out}	outer cylinder radius (m)
S	strain energy (J)
S_{max}	maximum strain energy (J)
t	time (s)
t_e	elastic time constant (s)
t_p	process time (s)
t_r	response time of the drop (s)
u	velocity (ms^{-1})
u_m	mean fluid velocity ($m\,s^{-1}$)

u_{max} maximum fluid velocity ($m\,s^{-1}$)
u_s surface velocity ($m\,s^{-1}$)
Δu_s surface velocity difference ($m\,s^{-1}$)
u_x velocity in x direction ($m\,s^{-1}$)
u_y velocity in y direction ($m\,s^{-1}$)
V volume (m^3)
W geometrical constant (–)
y y-co-ordinate (–)
α approach angle (°)
ε energy dissipation rate ($W\,kg^{-1}$)
ε_m mean energy dissipation rate ($W\,kg^{-1}$)
ε_{max} maximum energy dissipation rate ($W\,kg^{-1}$)
ϕ phase angle (°)
γ shear rate (s^{-1})
γ_{avg} average shear rate (s^{-1})
γ_n normal shear rate (s^{-1})
λ constant (–)
λ_k Kolomogoroff microscale of turbulence (m)
μ viscosity (Pa s)
μ_{con} viscosity of continuous phase (Pa s)
μ_{dis} viscosity of dispersed phase (Pa s)
ν kinematic viscosity ($m^2\,s^{-1}$)
θ rotational co-ordinate (rad)
ϱ density ($kg\,m^{-3}$)
σ stain (s^{-1})
σ_i additional strain (s^{-1})
σ_{max} maximum elastic strain (s^{-1})
σ_o time independent strain (s^{-1})
σ_T surface tension (Pa)
τ stress ($N\,m^{-2}$)
τ_{90} shear stress at the equator of the cell ($N\,m^{-2}$)
τ_{avg} average shear stress ($N\,m^{-2}$)
τ_{elong} elongational stress ($N\,m^{-2}$)
τ_{ent} stress at the entrance ($N\,m^{-2}$)
τ_m mean stress ($N\,m^{-2}$)
τ_{max} maximum stress ($N\,m^{-2}$)
τ_n normal stress ($N\,m^{-2}$)
$\tau_{p\text{-}f}$ interaction shear stress between particle and fluid ($N\,m^{-2}$)
ω angular velocity (s^{-1})
ω_{crit} critical angular velocity (s^{-1})
ω_o vorticity parameter (s^{-1})

1 Introduction

In the bioprocess industries materials that are adversely affected by their engineering flow environment are loosely and collectively referred to as "shear

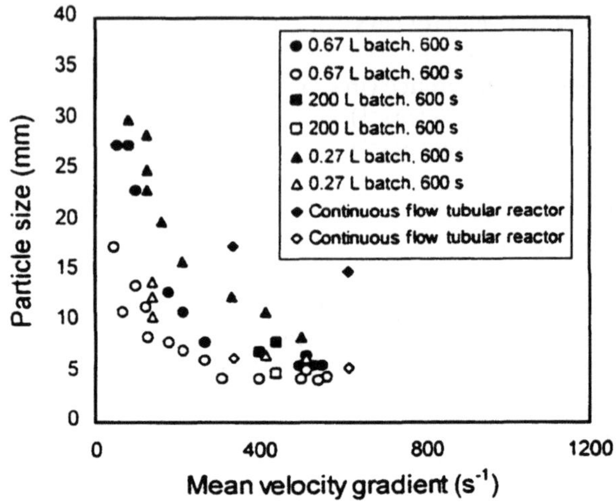

Fig. 1. Relationship between reactor mean velocity gradient and particle size for isoelectric soya protein precipitate. *Open symbols* represent precipitate diameter for 50% oversize; *Closed symbols* represent precipitate diameter for 90% oversize [51]

sensitive". The deleterious physical and biological effects of flow have been documented for a wide and varied range of freely suspended and anchorage-dependent biomaterials including mammalian, insect, microbial and plant cells [1–17], macromolecules such as proteins and DNA in solution [18–35], antibody fragments [36, 37], nematodes [38] and human tissue cells [39]. These effects have been confirmed at plant operations with bioprocess equipment such as high pressure homogenisers [40], centrifuges [8, 41], bioreactors (Fig. 1) [23, 42, 43] and pumps (Fig. 2) [24, 42, 44–46] and in laboratory batch and continuous flow devices [23, 44–59]. Despite numerous publications on the subject however no consensus of opinion has been reached on the precise mechanisms by which flow induces the observed effects. Important generic conclusions that have emerged from these studies include the following:

(a) Physical and biological responses of biomaterials are strong functions of the flow strength which is often expressed in terms of a local or mean bulk fluid stress or an energy dissipation rate. Additionally, these responses appear to depend on changes in the imposed stress as a function of time. For example, experimental evidence suggests that bioparticles in suspension are often more sensitive to rapidly time-varying, fluctuating Reynolds stresses in turbulent flow than steady stresses of an equal average value [8, 9, 16, 60].
(b) Most studies assume that shear stresses originating from laminar and turbulent flow are responsible for the observed effects. In several studies however the elongation flow has also been shown to induce similar effects [30, 35, 61, 62].
(c) For a given flow strength, the nature and the extent of the responses are determined by the biomechanical and rheomechanical properties of the materials which are poorly understood [63].

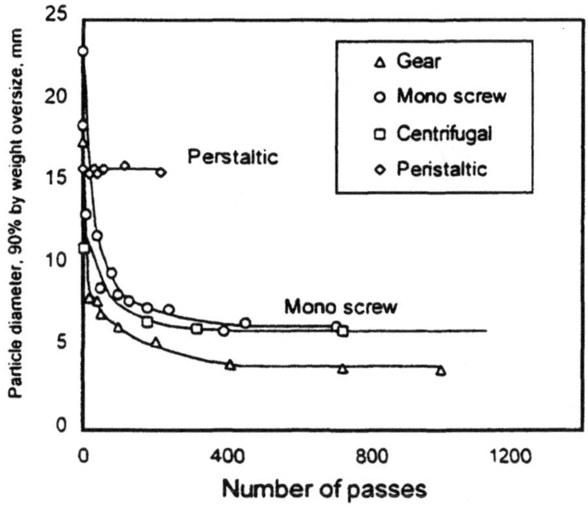

Fig. 2. Flow of protein precipitates through various pumps [107]

Qualitatively, the evidence suggests that the outer walls of most flow sensitive biomaterials consist of a biomechanically ill-defined cellular and interstitial network of biopolymers. Physiological activity at the wall results in significant heterogeneity in the material properties within the network and according to the classical theories dynamic changes of the network structure can be induced by external stresses such as those developed by the local fluid flow. A picture that emerges from these observations is of a biocolloid which possesses strong viscoelastic properties [64]. The load bearing capacity of the walls of biomaterials is rarely studied but many have been shown to stretch and elongate considerably under stress. The elastic behaviour of red blood cells for example is well documented [63, 65–73] and contributes significantly to their ability to resist flow induced deformations. The peptidoglycan, the major biopolymer constituent of cell walls of bacteria has been reported to be capable of stretching by as much as 400% when stressed [74] and DNA strands can double their length before breaking under an imposed flow stress [35].

It is likely that most biomaterials possess non-linear elastic properties. However, in the absence of detailed measurements of the relevant properties it is not necessary to resort to complicated non-linear theories of viscoelasticity. A simple dashpot-and-spring Maxwell model of viscoelasticity will provide a good basis to consider the main features of the behaviour of the soft-solid walls of most biomaterials in the flow field of a typical bioprocess equipment.

A Maxwellian material can be described by the following constitutive equation:

$$\frac{du}{dy} = \frac{\tau}{\mu} + \left[\frac{d\tau}{dt}\right]\left[\frac{1}{G}\right] \tag{1}$$

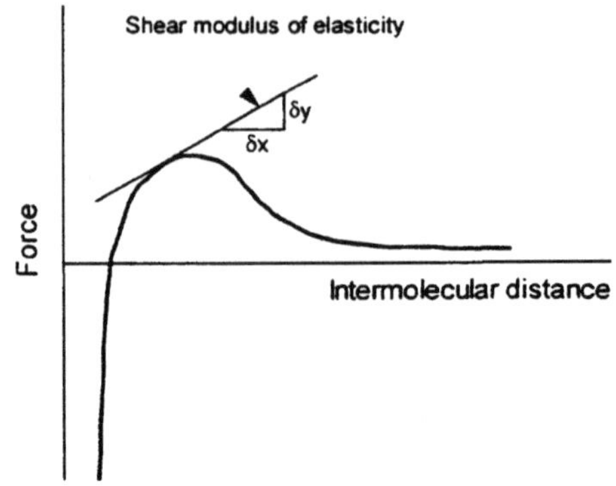

Fig. 3. Force-distance and energy-distance plots

overall wall deformation caused by an imposed stress, τ. The term (dτ/dt) represents the time change in that stress and G is the shear modulus of elasticity and gives a measure of the elastic effects. Likewise, μ is the shear viscosity.

Multiplying both sides of Eq. (1) by μ gives:

$$\mu \frac{du}{dy} = \tau + \left[\frac{d\tau}{dt}\right]\left[\frac{\mu}{G}\right] \tag{2}$$

The ratio (μ/G) has the units of time and is known as the elastic time constant, t_e, of the material. Little information exists in the published literature on the rheomechanical parameters, μ, and G for biomaterials. An exception is red blood cells for which the shear modulus of elasticity and viscosity have been measured by using micro-pipette techniques [66, 68, 70, 72]. The shear modulus of elasticity data is usually given in units of $N\,m^{-2}$ and is sometimes compared with the interfacial tension of liquids. However, these properties are not the same. Interfacial tension originates from an imbalance of surface forces whereas the shear modulus of elasticity is an interaction force closely related to the slope of the force-distance plot (Fig. 3). Typical reported values of the shear modulus of elasticity and viscosity of red blood cells are $6 \times 10^{-5}\,N\,m^{-1}$ and 10^{-3} Pa s respectively [70]. Red blood cells typically have a mean length scale of the order of 7 μm, thus G is of the order of $10\,N\,m^{-2}$ and the elastic time constant (μ/G) is of the order of 10^{-4} s.

The ratio of a material's elastic time constant, t_e, to the time of an imposed process, t_p, is the dimensionless Deborah number, De $(=t_e/t_p)$. When the imposed process time, t_p, is rapid, the De number will be high, elastic stresses will dominate, the first (viscous) term in Eq. (2) will be negligible and the material will behave purely elastically. For rapid processing operations therefore, Eqs. (1) and (2) can be rewritten as:

$$\frac{du_y}{dy} = \left[\frac{d\tau}{dt}\right]\left[\frac{1}{G}\right] \tag{3}$$

where the deformation is now taken to be in the direction of flow, i.e. elongation (stretching) flow. Assuming that the local stress changes from 0 to τ ($=\tau_{elong}$) over a short period of process time, t_p, Eq. (3) can be rearranged to give approximately:

$$\left[\frac{du_y}{dy}\right] = \frac{\tau_{elong}}{t_p}\frac{1}{G} \tag{4}$$

The engineering significance of Eq. (4) can be seen, for example, by considering the bioprocess operation of disruption of unicellular micro-organisms in a high pressure homogeniser (HPH). Typical values of imposed stress, τ, and process time, t_p, in the chamber of a HPH are 10^8 N m^{-2} and 10^{-6} s respectively [40]. The imposed stress is achieved rapidly and process time, t_p, can be assumed to be close to the residence time in the chamber. The rate of change of stress therefore is of the order of 10^{14} N m^{-2} s^{-1}. Taking a mammalian (red blood) cell as an example and using a value of 10 N m^{-2} for the shear modulus of elasticity, G, in Eq. (4) gives an elongation deformation rate of the order of 10^{13} s^{-1} which is significantly higher than the typical shear rate in the same unit which is of the order of 10^7 s^{-1}. The response of the cells to such a strong elongation flow is expected to be highly elastic. Models based on elongation stresses have been used successfully in the past to describe the breakage of yeast cells [40] and liquid drops [75, 76] in high pressure homogenisers.

Returning to Eq. (1) and considering the response of a biomaterials to an imposed stress in a relatively slow process (long process time, t_p and low Deborah number, De), the elastic component, the second term, of Eq. (1) will be small and the wall material is likely to behave in a viscous manner. Thus Eq. (1) reduces to:

$$\tau = \mu \frac{du_g}{dy} \tag{5}$$

where τ and (du_x/dy) are the shear stress and shear rate respectively.

First order calculations show that in many items of bioprocess equipment, flow fields are often such that both terms in Eq. (1) are significant. The preceding discussion suggests that for an imposed flow the biomechanical wall properties determine the nature and the extent of the biological and physical responses of freely suspended bioparticles. The present review of the published literature was undertaken in order to assess the current state of knowledge and to identify gaps in our understanding of the key flow associated parameters that adversely affect such materials. In this respect it ought to be mentioned that first, biological effects of flow are not always harmful. There are reported cases where a relatively low level of fluid induced stress has been shown to have beneficial effects on, for example, cell physiology [8, 77]. However, in cases where such effects are detrimental they severely limit the choice of process options and complicate the design of process equipment. Secondly, physiological effects of flow are often observed at relatively low to moderate levels of stress (0.5–5.0 N m^{-2}) and do not always require as a prerequisite the material to physically break. For stresses larger than about 5 $N m^{-2}$ the mechanical network forming the external walls undergo permanent structural and irrecoverable damage. Under such conditions, walls of many unicellular micro-organisms will rupture, the backbone of DNA strands will fracture, and the individual hyphae will break off the mycelium in microbial cultures.

2
Rheomechanical Properties of Biomaterials

Figures 4 and 5 give a broad indication of the relevant biomechanical properties of a number of flow sensitive biomaterials. In the case of the data shown in Fig. 5, the surface mechanical properties are lumped into a single measure of the surface integrity. Admittedly, in view of what has been said in the introduction about the viscoelastic nature of the wall material, the information given in Figs. 4 and 5 are oversimplistic. The data in Fig. 5 are based on reported critical minimum stresses (often expressed in terms of the mean bulk fluid stresses) at which physical damage is first observed. Figure 6 gives an indication of the

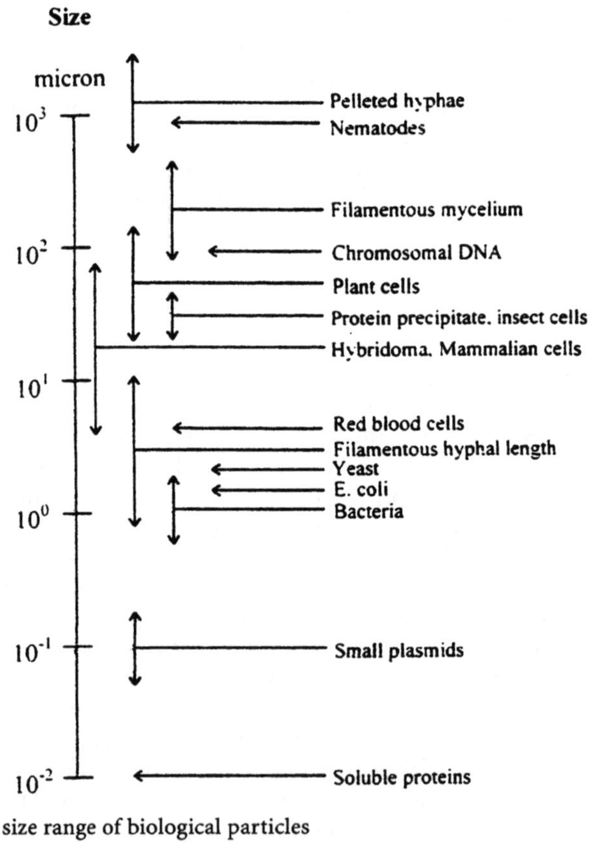

Fig. 4. Typical size range of biological particles

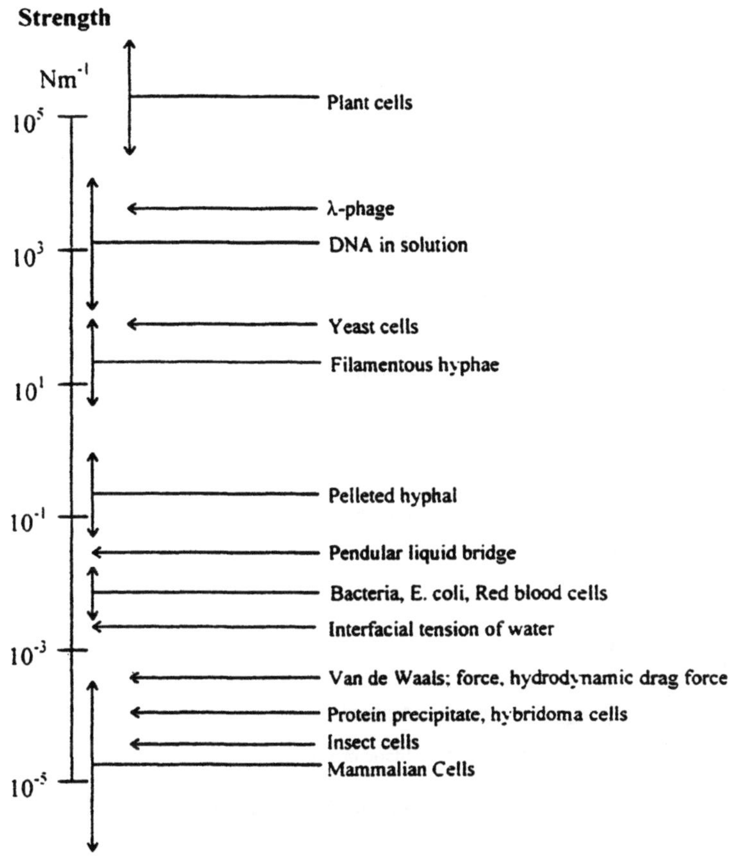

Fig. 5. Estimated characteristic strength of typical biological particles of interest to biotechnology data are based on in-situ measurements of the minimum stresses necessary to cause permanent breakage of particles. For comparison data are shown based on Van der Waals' and pendular liquid bridges between two 10-μm particles, 0.01 μm apart

mean energy dissipation rate in several key items of bioprocess equipment. In Figs. 4–6 the uncertainty that exists in using absolute values for the constants in the equations results in large variations in the predictions of, for example, surface strength and energy dissipation rate. These figures therefore should be treated with caution and are included only to serve as a general basis for discussion.

The development of the biophysical processes that lead to the mechanical strength of the deformable walls of micro-organisms is not well understood. Limited work done with filamentous fungi has shown that complex time-dependent biopolymerisation reactions occur at the wall throughout the growth cycle. These studies suggest that extrusion at the hyphal tip and subsequent time dependent polymerisation of the excreted precursors and enzymes gradually lead to the formation of a complex mechanical network with a high

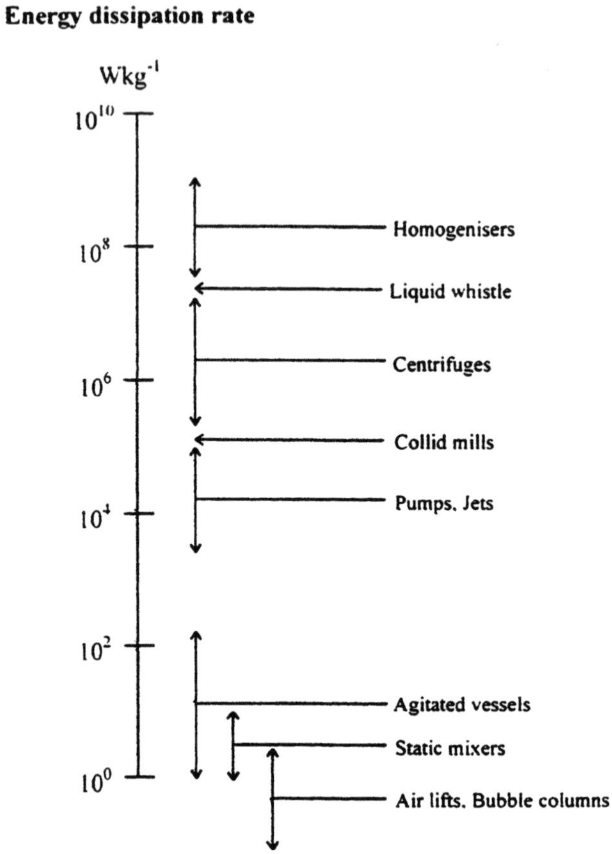

Fig. 6. Estimates of fluid energy dissipation rates in different items of bioprocess equipment

degree of cross-linking. The gradual biopolymerisation reaction causes the wall thickness to increase with distance away from the growing tip. For example, for *N. crassa*, measurements have shown that the wall thickness increases from 0.05 µm to about 0.15 µm at a distance of about 300 µm away from the tip. Similar observations involving biopolymerisation reaction at the wall and an increase in cell wall thickness with cycle time have been reported for unicellular micro-organisms including mammalian and bacteria cells [63, 67, 73].

The variation in wall thickness and the development of cell wall rigidity (stiffness) with time have significant consequences when considering the flow sensitivity of biomaterials in suspension. For an elastic material, stiffness can be characterised by an elastic constant, for example, by Young's modulus of elasticity (E) or shear modulus of elasticity (G). For a material that obeys Hooke's law, for example, a simple linear relationship exists between stress, τ, and strain, σ, and the ratio of the two uniquely determines the value of the Young's modulus of the material. Furthermore, the (strain) energy associated with elastic deformation of an object with volume, V, when Hooke's law is applicable, is given in most textbooks as the strain energy:

$$S = \frac{V\tau\sigma}{2} = \frac{V\tau^2}{2E} \tag{6}$$

For a Hookian material, the concept of minimum strain energy states that a material fails, for example cell wall disruption occurs, when the total strain energy per unit volume attains a critical value. Such an approach has been used in the past to describe a number of experimental observations on the breakage of filamentous micro-organisms [78, 79]. Unfortunately, little direct experimental data are available on the Young's modulus of elasticity, E, or shear modulus of elasticity G representing the wall properties of biomaterial. Few (natural) materials behave in an ideal Hookian manner and in the absence of any other information, it is not unreasonable to assume that the mechanical properties of the external walls of biomaterials will be anisotropic and anelastic.

Elastic anisotropy (different elastic constants in different orientations) can have profound effects on the behaviour of biomaterials in suspension. For example, consider a single cell micro-organism in a uniform flow field with a constant strength which is not so excessive as to cause the complete break-up of the cell. If the cell walls have significantly different elastic constants in different orientations, then stress heterogeneity will occur within the walls network with the effect that breakage becomes dependent on cell orientation in the flow field. The cells will display a degree of time-dependent response to the imposed fluid stress since not all the cells will be aligned in the orientation of least resistance to the flow at the same time.

This apparent time dependent cell disruption is caused because of the statistically random distribution of the orientation of the cells within a flow field and the random changes in that distribution as a function of time, the latter is caused as the cells spin in the flow field in response to the forces that act on them. In the present discussion this is referred to as apparent time dependency in order to distinguish it from true time-dependent disruption arising from anelastic behaviour of the cell walls. Anelastic behaviour, or time-dependent elasticity, is thought to arise from a restructuring of the fabric of the cell wall material at a molecular level. Anelasticity is stress induced and requires energy which is dissipated as heat, and if it is excessive it can weaken the structure and cause its breakage.

The effects of anelasticity are seen clearly in the strain-time schematic plot of Fig. 7. The time independent strain, σ_o, occurs immediately as the stress is applied. The additional strain, σ_i, is associated with stress-induced anelasticity. The total strain, $\sigma_o + \sigma_i$, is approached exponentially. When the stress is removed the time independent strain is recovered immediately, while the anelastic component relaxes exponentially with a relaxation time constant, k.

Consider the case in which the imposed stress changes cyclically according to the following equation:

$$\tau = \tau_{max} \sin \omega t \tag{7}$$

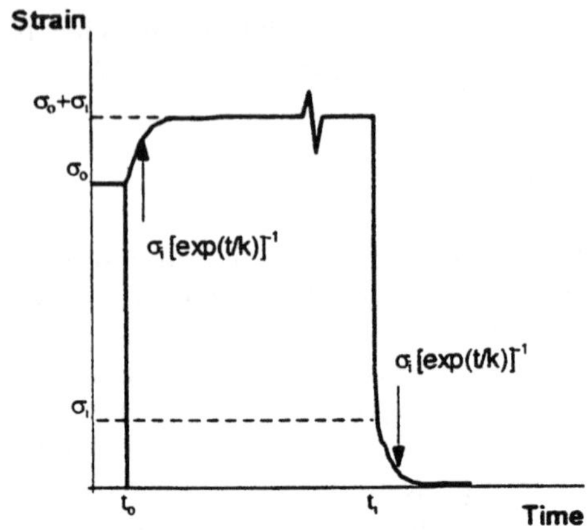

Fig. 7. Strain-time plot

where τ_{max} is the maximum stress, τ is the stress at time, t, ω is the angular velocity of stress cycling. For a perfectly elastic body the strain is in phase with the stress and is given by:

$$\sigma = \sigma_{max} \sin \omega t \tag{8}$$

If the material is anelastic, the stress and the strain will not coincide. The strain will lag behind by an amount which is determined by the phase angle, ϕ. Thus:

$$\sigma = \sigma_{max} \sin(\omega t - \phi) \tag{9}$$

The result is a hysteresis loop on a stress-strain plot (Fig. 8). The energy loss per cycle is the area of the hysteresis loop and is given by:

$$\int_0^{2\pi} \tau \, d\sigma = \int_0^{2\pi} \tau_{max} \sigma_{max} \sin \omega t \, d(\omega t - \phi) = \pi \tau_{max} \sigma_{max} \sin \phi \tag{10}$$

Noting that the maximum elastic strain energy:

$$S_{max} = \frac{\tau_{max} \sigma_{max}}{2} \tag{11}$$

each cycle therefore dissipates $2\pi \sin \phi$ as the fractional energy loss. The effects of this loss of energy within the cell wall structure of biomaterials have not been reported in the past but are perceived to have important engineering implications. One consequence of the anelastic energy dissipation, $2\pi \sin \phi$, is a reduction in the total strain energy of the cell wall material and hence the critical stress required to cause disruption will appear to increase. Hence, for otherwise

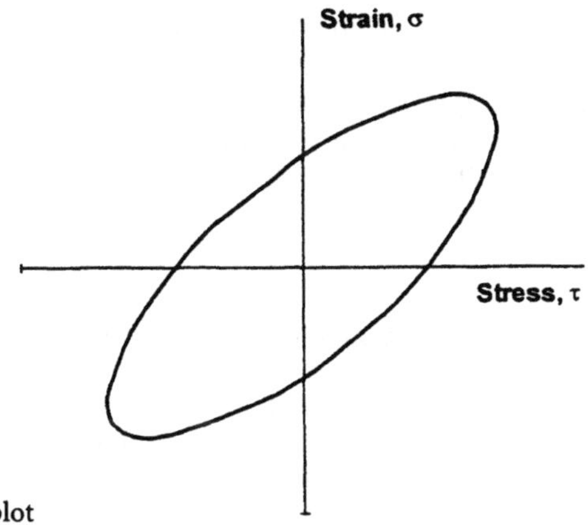

Fig. 8. Stress-strain plot

similar conditions physical disruption and possibly detrimental biological responses will start at higher stresses. It is often reported in the literature that many biological materials have the ability to adapt to their hydrodynamic environment. The improved ability for a cell to resist disruption can be a direct consequence of the reordering of the cell wall structure and the accompanying reduction in the total strain energy associated with anelasticity. Namdev and Dunlop [77] suggest that many organisms have the capacity to express certain "stress" proteins in response to environmental stresses. The "stress" proteins are perceived to provide a degree of protection to the cells against fluid stresses. To date however attempts to identify such proteins have been unsuccessful. Stress induced reordering of the cell wall structure associated with anelasticity is a physical process that does not require the expression of "stress" proteins. However, the energy dissipation associated with anelasticity is critically dependent on the stress cycling. If the strain lags the stress sufficiently so that there is an appreciable hysteresis loop as shown in Fig. 8, the total energy dissipation can be large with rapid stressing (many cycles per second). This energy loss shows up as heat within the body and can weaken the structure.

In addition, flow deformation can distort the shape of many micro-organisms in suspension. Complications arise because the change in shape, for example the elongation of a spherical cell into a cylindrical body or its flattening into a disc shape object, will alter its projected area exposed to a flow stress. For example, recent experimental evidence with endothelial cells and numerical simulations [80] with mammalian cells in laminar flow have shown that, for a fluid stress of 2.5 $N\,m^{-2}$ the (drag) force on the cells changes by 50% and the associated torque by 95% as a consequence of changes in the projected area. The situations are further complicated if the fluid stresses fluctuate in intensity, as they would do for example in a turbulent flow. In the next section a number of possible flow fields of interest are reviewed and their effects on freely dispersed biomaterials discussed. Since at present no method of analysis is available to account for the combined effects of wall anelasticity and flow heterogeneity on deformation, the following discussion assumes that the mechanical properties of biomaterials are given by a single, time independent strength parameter (Fig. 5). Moreover, in the absence of relevant information, the discussion is simplified by neglecting distortions in the shape of the micro-organisms. Although these assumptions necessarily limit the scope of the discussion, the information that it provides is thought to be of considerable general interest.

3 Flow Stresses Acting on Biomaterials in Suspension

Considerable work has been done in the past in formulating mechanistic models of the physical processes that are perceived to be responsible for the deleterious effects of flow. In practically all of these a primary flow stress is identified as the cause of the observed effects. The most relevant flow stresses are thought to originate from:

(a) Particle-particle collisions in which the energy of impact is derived from the motion of the surrounding fluid which is often assumed to be turbulent;
(b) Impacts between bioparticles and other inserts within the equipment, particle motion and its kinetic energy are normally based on fluid turbulent motion;
(c) Turbulent shear and normal (elongation) stresses arising from the local turbulent flow field in the equipment;
(d) Laminar shear and elongation stresses arising from local velocity gradients at a solid boundary and/or in well-defined flow fields;
(e) Physical and physico-chemical interactions between the bioparticles and gas (bubbles) interfaces in the presence and/or absence of metal ions and contaminants.

3.1 Turbulent Flow Stresses Acting on Biomaterials

Stresses acting on micro-organisms in (a) to (c) are derived on the premise that the flow forces originate from the turbulent motion of the carrier medium. In almost all cases, turbulence is assumed to be locally isotropic and homogeneous which greatly simplifies the analysis and allows the application of the Kolmogoroff's theory of turbulence to the problem [81]. The Kolomogoroff micro-scale of turbulence, λ_k:

$$\lambda_k = \left[\frac{\nu^3}{\varepsilon}\right]^{1/4} \tag{12}$$

determines the nature of the flow forces that act on the micro-organisms. Thus, for bioparticles with a size smaller than or comparable to the micro-scale of turbulence, shear forces arising from eddy motion in the viscous dissipation subrange of the turbulent energy spectrum are thought to be responsible for the flow effects. For organisms which are larger than the micro-scale of turbulence, eddies in the inertial subrange provide the main flow forces. In that case, the relevant forces are perceived to originate from fluctuating surface pressure resulting in normal (elongation) stresses.

The basic application of these concepts has been firmly established over the past two decades for both biological and non-biological suspensions. Analysis of flow damage based on the concept of isotropic turbulence suggests that stresses originating from (a) and (b) are unlikely to be important because most biomaterials of interest are relatively small and have densities only marginally different from those of the carrier fluid. Thus, inertia forces tend to be negligible, unless the material is anchored onto the surface of an inert (large and heavy) particle. This is sometimes done with exceptionally fragile cells in order to enhance their physiological characteristics in suspension. Turbulent flow breakage of anchorage dependent cells has been examined and reviewed extensively elsewhere [82–84].

For freely suspended bioparticles the most likely flow stresses are perceived to be either shear or normal (elongation) stresses caused by the local turbulent flow. In each case, there are a number of ways of describing mathematically the interactions between turbulent eddies and the suspended particles. Most methods however predict the same functional relationship between the prevailing turbulent flow stresses, material properties and equipment parameters, the only difference between them being the constant of proportionality in the equations. Typically, in the viscous dissipation subrange, theory suggests the following relationship for the mean stress [85]:

$$\tau_m = \left[\frac{2\mu}{15}\left(\frac{\varepsilon}{\nu}\right)\right]^{1/2} = \left[\frac{2\varrho}{15}(\varepsilon\nu)\right]^{1/2} \tag{13}$$

where in the inertial subrange the mean stress is given by:

$$\tau_m = 1.88\,\varrho\,[\varepsilon d]^{2/3} \tag{14}$$

Turbulent disruption of biocolloids, such as single cell micro-organisms with average sizes less than the micro-scale of turbulence, was recently analysed by Cherry and Kwon [86] by assuming that disruption occurs by flow stresses inside the eddy. Micro-scale eddies are pictured as long, flexible rotating cylinders which are in a constant state of deformation induced by larger scale flows. The deformation of an eddy causes it to stretch axially producing a radially inward flow within it. Small spherical bioparticles such as unicellular micro-organisms are able to move in and out of these eddies. Once inside an eddy the micro-organism will experience a distribution of shear rates and shear stresses resulting from the velocity profiles within it. Flow approximations can be made and the resulting equations solved to obtain the fluid shear rate and cell surface shear stress distributions. The final expressions are given below:

$$\tau = \frac{5}{2}\,\mu\gamma\sin\theta\,[1 - \sin^2(\gamma t)\sin^2\theta]^{1/2} \tag{15}$$

where the fluid shear rate, γ, is given by:

$$\gamma = \left[\frac{-4\nu\,\omega_o}{ar^2}\right] + \left[\frac{4\nu\,\omega_o}{ar^2} + \omega_o\right] \exp\left[\frac{-ar^2}{4\nu}\right] \tag{16}$$

$$= \omega_o \left[-\frac{1}{R} + \left(\frac{1}{R} + 1\right)\right] \exp(-R) \tag{17}$$

where

$$a = 0.18\left[\frac{\varepsilon}{\nu}\right]^{\frac{1}{2}};\ \omega_o = 7.14\left[\frac{\varepsilon}{\nu}\right]^{\frac{1}{2}};\ r = r_o \exp\left[-\frac{at}{2}\right] \text{ and } R = \left[\frac{ar^2}{4\nu}\right] \tag{18}$$

where a is the strain rate parameter, ε is the turbulent energy dissipation and ω_o is the vorticity parameter. The constant r_o is an arbitrary starting point of the cell's position in the eddy at time t = 0.

Equation (15) indicates that the wall shear stress changes with position on the cell surface and at each position the stress changes with time. Figures 9 and 10 show the distributions of the fluid shear rate in the eddy and the cell surface shear stress as a function of the dimensionless radius, R, and time, γt, respectively.

In Eq. (15), θ = 90° represents the equator of the cell where shear stress has its maximum distribution, given by:

$$\tau_{90} = \frac{5}{2}\mu\gamma(\cos\gamma t) \tag{19}$$

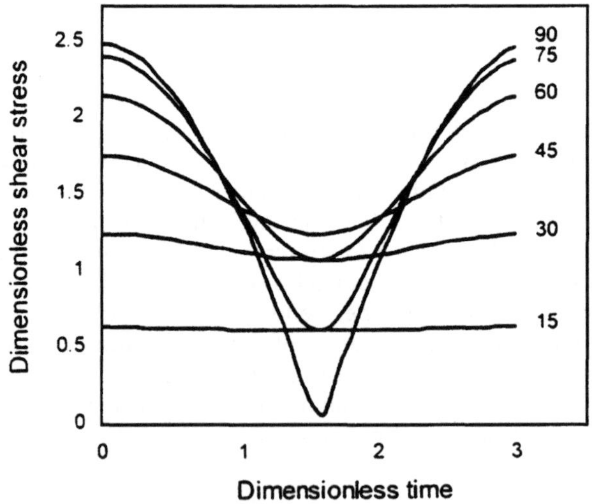

Fig. 9. Dimensionless shear stress τ/μγ as a function of θ and dimensionless time γt [86]

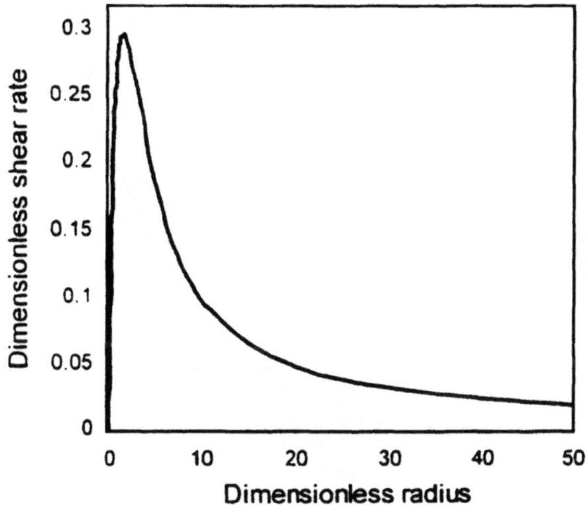

Fig. 10. Local values of the dimensionless shear rate in the model eddy [86]

The maximum fluid shear rate in the eddy occurs at the radial position R = 1.793 [86] (Fig. 10). Thus, letting $\cos \gamma t = 1.0$, in Eq. (19) gives the maximum shear stress acting on the cell as:

$$\tau_{90}\big|_{R=1.793} = 5.33\,\mu \left[\frac{\varepsilon}{\nu}\right]^{1/3} = 5.33\,\varrho\,(\varepsilon\nu)^{1/3} \tag{20}$$

3.2 Flow Stresses Acting on Particles in Bioprocess Equipment

In all of the above equations, ε is assumed to be constant and uniform throughout the flow field. In most items of bioprocess equipment, however, there is a spatial distribution of energy dissipation. The definition of an average or a maximum energy dissipation rate is notoriously difficult in the case of bioprocess equipment such as high pressure homogenisers, centrifuges, pumps and microfiltration units which all have complex flow fields.

3.2.1 *Tubular Bioreactors*

Numerous researchers have studied damage to micro-organisms during flow in pipes, (Fig. 11) [87, 88] Most researchers use a Fanning friction factor, f, to calculate the energy dissipation rate for fully developed flow in tubular bioreactors and capillary flow devices. There are minor differences in the equations that are used but they are generally of the following form [89, 90]:

$$\frac{\Delta P}{L_p} = \frac{4}{D}\frac{\varrho u^2 f}{2} = \frac{32 \varrho Q^2 f}{\pi^2 D^5} \tag{21}$$

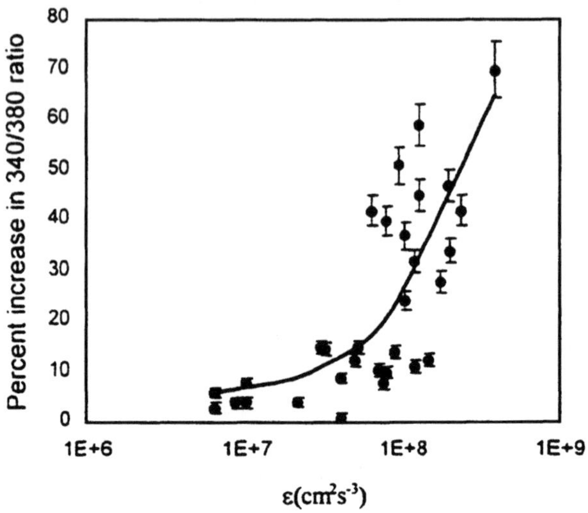

Fig. 11. Capillary data showing an increase in the release of intracellular calcium content (340/380 ratio) as ε increases. Data refer to experiments in which the cell suspension was circulated through a fine bore capillary for 15 min. *Error bars* represent the 8.0% standard deviation [87]

$$\text{where}\quad f = \frac{16}{Re} \quad \text{for} \quad Re < 2000$$

$$f = \frac{0.08}{Re^{0.25}} \quad \text{for} \quad 2000 < Re < 4000 \tag{22}$$

$$f = 0.001 \quad \text{for} \quad Re > 3 \times 10^5$$

Dividing Eq. (21) by the density and multiplying by the velocity gives the energy dissipation rate per unit mass. Thus:

$$\varepsilon = \frac{\Delta P}{L_p}\frac{u}{\varrho} = \frac{128\,Q^3 f}{\pi^3 D^7} \tag{23}$$

For Re < 2000, i.e. laminar flow conditions, using Eq. (22) to replace f in Eq. (23) gives:

$$\varepsilon = \frac{512\,\mu Q^2}{\pi^2 \varrho D^6} \tag{24}$$

Similarly, for transition flow, 2000 < Re < 4000:

$$\varepsilon = \frac{7.24\,\mu^{0.25} Q^{2.75}}{\pi^{2.75} \varrho^{1.25} D^{6.75}} \tag{25}$$

The application of Eqs. (21) to (25) assumes fully developed velocity profiles. The length of pipe, L_p, over which the velocity develops is a function of the operating Reynolds number. An estimate of L_p can be made by using the following equation:

$$\frac{L_p}{D} = 1.41\,(Re)^{0.25} \tag{26}$$

Equation (26) gives values of L_p equal to 10–16 pipe diameters for Re between 2500 and 20000. According to Davies [91] these are the lowest estimate of the hydraulic entry length. In practice the observed values are often in the range of 25 to 100 pipe diameters.

In the developing flow region, Zhang et al. [88] recommend the use of an apparent friction factor defined as:

$$f_{app} = \frac{0.296}{\left[Re_p\left(\frac{L_p}{D}\right)\right]^{0.2}} \tag{27}$$

and

$$\Delta P = \frac{1}{2} f_{app}\,\varrho \left[\frac{L_p}{D}\right] u_{max}^2 \tag{28}$$

where the maximum velocity at the centre of the pipe, u_{max}, is given by:

$$u_{max} = \frac{u_m}{0.817} \tag{29}$$

where u_m is the mean velocity.

3.2.2
Mechanically Stirred Bioreactors

Figures 12 and 13 show the effects of agitation and time of exposure on suspensions of biological materials in bioreactors [61]. In turbulent flow the energy dissipation rate per unit mass, ε, of a stirred bioreactor is normally expressed by the following equation:

$$\varepsilon = \frac{P}{\varrho V} \tag{30}$$

where P is the impeller power input and is given by:

$$P = K\varrho N^3 D_{imp}^5 \tag{31}$$

where D_{imp} is impeller diameter and K is an experimentally determined constant which depends on the type of impeller. For a typical six bladed disc impeller, for example, K has a value between 5 and 6 [92].

$$\varepsilon = \frac{KN^3 D_{imp}^5}{V} \tag{32}$$

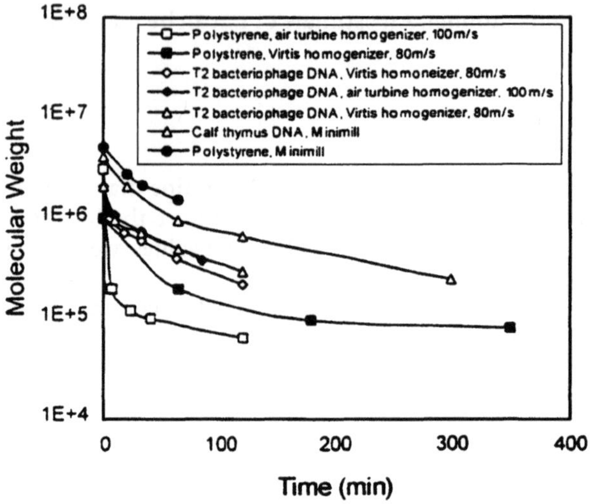

Fig. 12. Molecular weight falloff with time for various rotating-blade shearing devices [32]

Using the total volume of the contents in the tank to calculate V in Eq. (32) gives the mean energy dissipation rate. However, significant flow heterogeneity exists in stirred vessels [93–96] which causes considerable uncertainties in the calculation of the energy dissipation rate, ε. Several researchers have argued that the calculation of ε ought to be based on the high energy density region around the impeller blade where damage to bioparticles is most likely to occur and where turbulence can be assumed to be approximately isotropic. For example, for standard Rushton turbine impellers in geometrically similar systems, the high energy dissipation volume is given by several equations in terms of the impeller diameter. These include $\varepsilon = 1.24\,(D_{imp})^3$ [97], $\varepsilon = 1.0\,(D_{imp})^3$ [94], $\varepsilon = 0.17\,(D_{imp})^3$ [98] and $\varepsilon = 0.075\,(D_{imp})^3$ [99] depending on the definition of the high energy region around the impeller.

A survey of the published literature indicates that the ratio of the maximum to mean energy dissipation rate in the vessel, $\varepsilon_{max}/\varepsilon_m$ can vary substantially but typically in the range 10 to 100 [85]. Recent measurements [100] of the turbulent flow properties with a range of impellers and vessel configurations indicate that the differences between the reported ratios of $\varepsilon_{max}/\varepsilon_m$ are partly due to differences in the geometrical variables. For example, detailed factorial designs of experiments showed significant effects of impeller diameter to tank diameter ratio and off-bottom clearance to impeller diameter ratio on the value of $\varepsilon_{max}/\varepsilon_m$.

The uncertainty involved in the calculation of the energy dissipation rate makes it difficult to compare experimental results reported by different researchers. For the same reasons, so far it has proved difficult to assess flow induced effects in different items of process equipment using the common basis of equal energy dissipation rate. For example, Fig. 14 shows the biological response of (SF-9) insect cells as a function of the energy dissipation rates in a capillary tube and in a mechanically stirred vessel [99]. In these plots the calcula-

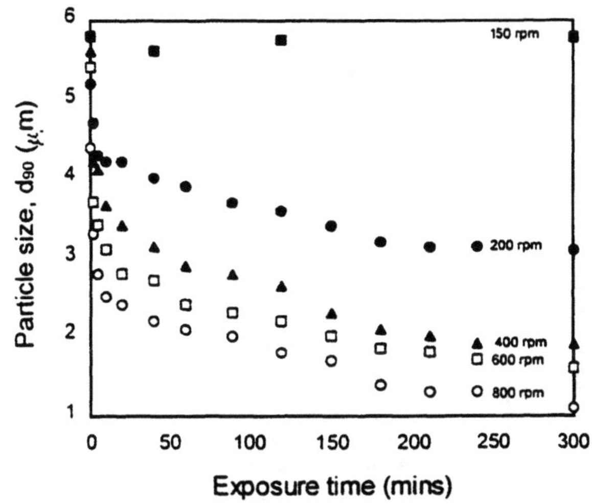

Fig. 13. The plot of d_{90} against time of exposure with impeller speed as a parameter. Data are for 300 min of exposure, with a 0.12 m diameter impeller and a protein concentration of 3.5 kg m^{-3}. Data are for five separate experiments, all involving the same aging conditions [61]

tions of energy dissipation rates in the tank assume the dissipation volume to be equal to 0.075 D_{imp}^3 whereas the estimations of the energy dissipation rates in the capillary are based on the mean velocity. In other words, the maximum energy dissipation in the vessel is compared with the mean values in the capillary. It is possible that in the fine bore capillaries used in the experiments the maximum and the mean velocities were similar and this could explain the reason for similar biological responses observed in the two devices at equal energy dissipation rates (Fig. 14).

3.2.3
Concentric Cylinders

Flow fields generated in the gap between a pair of rotating cylinders have been used by a few researchers in the past to study flow associated damage to microorganisms (Figs. 15–19). Haas [101] recommended the following set of equations for the calculations of the mean energy dissipation rate per unit mass for flow between rotating cylinders:

$$\varepsilon_m = \frac{2\mu A u}{D_{in}} u \left[\frac{2}{D_{in} A \varrho}\right] = \frac{4\mu}{D_{in}^2 \varrho} u^2 \qquad (33)$$

where $u = \pi D_{in} N$ and D_{in} is the inner diameter of the rotating cylinder device. The definition of energy dissipation rate according to Eq. (33) allows a comparison of flow associated effects between a tubular bioreactor (Eq. (23)) and a rotating cylinder device. For laminar flow conditions, the fluid velocity will only

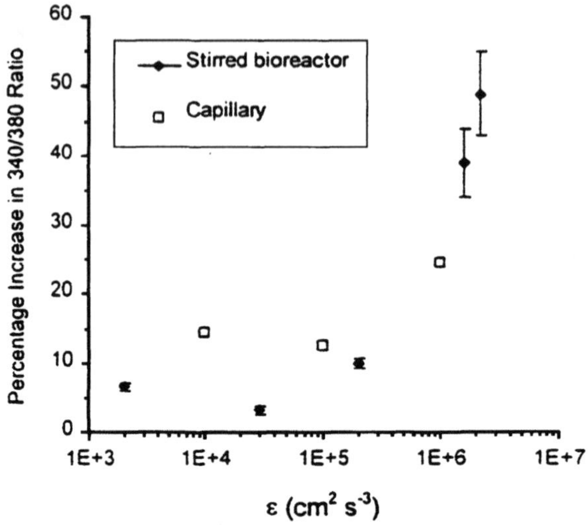

Fig. 14. Calcium response of Sf-9 insect cells subjected to different values of ε in a stirred bioreactor equipped with a 5.1 cm diameter 6-bladed Rushton impeller (*closed circles*) or in the capillary flow system (*open squares*). *Error bars* for stirred bioreactor are standard deviation for each experiment but for the capillary, data are hard to discern [99]

average half of the rotating velocity. Using Eq. (33) with:

$$Re = \frac{D_{in} \varrho u}{2\mu} \tag{34}$$

gives the energy dissipation rate for Couette flow expression, thus:

$$\varepsilon_m = \frac{4\mu}{D_{in}^2 \varrho} u = \frac{1}{8} \frac{fu^3}{D_{in}} \tag{35}$$

while for transitional flow:

$$\varepsilon_m = 0.01189 \frac{\mu^{0.25} u^{2.75}}{\varrho^{0.25} D_{in}^{1.25}} \tag{36}$$

However, several flow transition regimes have been identified between laminar and fully turbulent flow. The cessation of laminar Couette flow is marked by the appearance of Taylor vortices in the gap between the two cylinders. For the case of stationary outer cylinder, the critical angular velocity, ω_{crit}, of the inner cylinder at which these flow instabilities first appear can be estimated by using the following equations [102]:

$$\omega_{crit}^2 = \frac{\pi^4 \nu^2 (r_{in} + r_{out})}{2W (r_{out} - r_{in})^3 r_{in}} \tag{37}$$

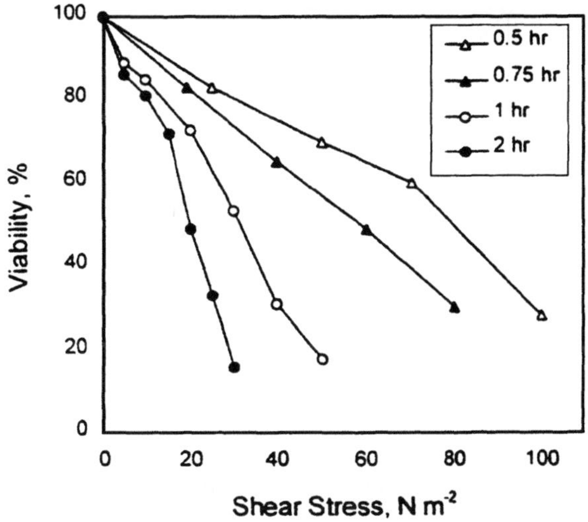

Fig. 15. Effects of turbulent shear stress level and exposure time on cell viability measured by trypan blue staining. Cells were sheared in a concentric cylinder viscometer [1]

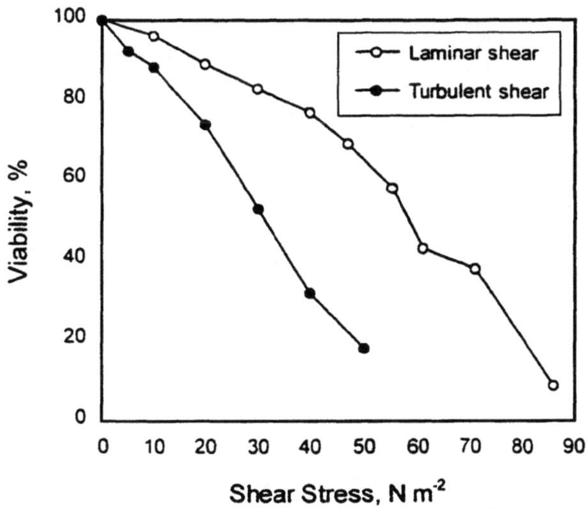

Fig. 16. Effects of flow regime on cell viability measured by trypan blue. Exposure time: 1 h [1]

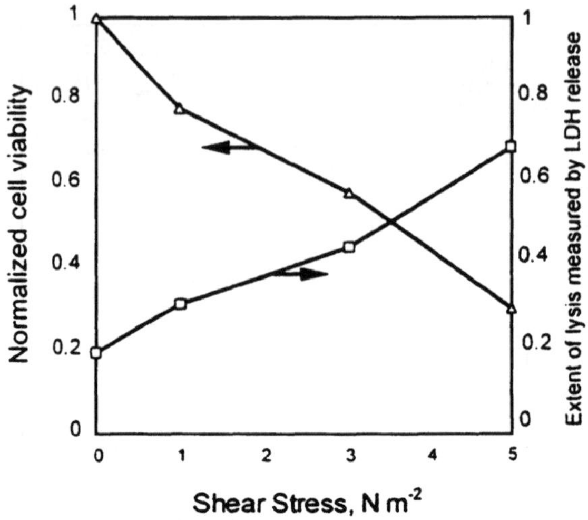

Fig. 17. Response of CRL-8018 hybridoma cells to increasing levels of well-defined laminar shear in the concentric cylinder viscometer for 10 min. Spinner flask cultures were seeded with cells from routine T-flask cultures that were 3 days old. Cell samples were taken from the spinner flask cultures during late-exponential growth and sheared in the viscometer [17]

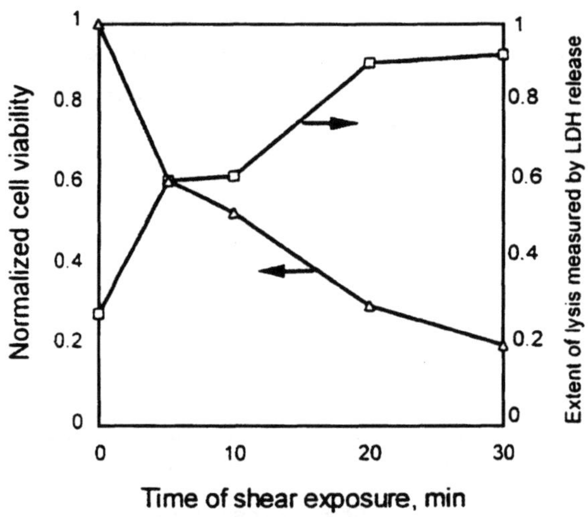

Fig. 18. Response of CRL-8018 hybridoma cells in well-defined laminar shear (50 dyne cm^{-2}) in the concentric cylinder viscometer. Spinner flask cultures were seeded with cells from routine T-flask cultures that were 3 days old. Cell samples were taken from the spinner flask cultures during late-exponential growth and sheared in the viscometer [17]

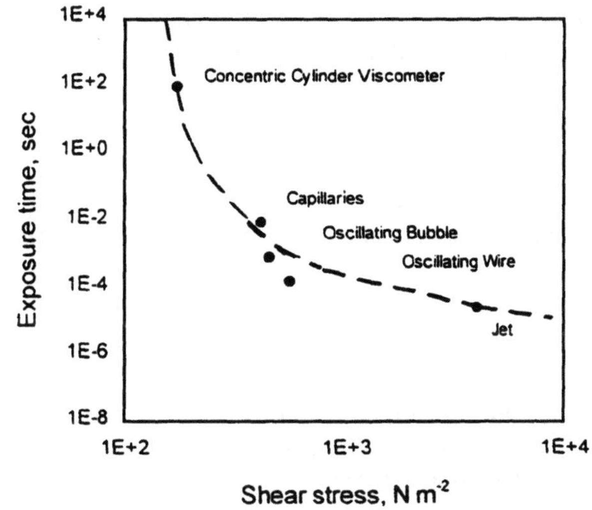

Fig. 19. Variation of exposure time associated with cell rupture and the threshold shear stress for various systems [112]

where

$$W = 0.0571\left[1 - 0.652\,\frac{r_{out} - r_{in}}{r_{in}}\right] + 0.00056\left[1 - 0.652\,\frac{r_{out} - r_{in}}{r_{in}}\right]^{-1} \quad (38)$$

where r_{in} and r_{out} are the radii of inner and outer cylindrical cylinders respectively.

3.2.4
Jet Flow Devices

A high velocity jet from a small diameter nozzle can produce exceptionally high energy intensity that can have profound effects on bioparticles (Fig. 19). Examples of bioprocess equipment in which flow can be closely approximated to that produced by a jet include pumps, impinging jet mixers and high pressure homogenisers (see Fig. 6). For a free jet the energy dissipation at the nozzle can be obtained from the following equation:

$$P = \frac{1}{2}\left[\frac{\pi D_{noz}^2}{4}\varrho u\right]u^2 \quad (39)$$

where D_{noz} is the nozzle diameter. Allowing for various energy losses leads to the following expression for the energy dissipation rate:

$$\varepsilon = C\,\frac{u^3}{D_{noz}} \quad (40)$$

with the empirical constant, $C = 0.1$. For a high pressure homogenisers, D_{noz} is usually assumed to be set equal to twice the gap width.

3.3 Laminar Flow Stresses Acting on Biomaterials

In a laminar shear flow, the maximum viscous shearing stress, τ_{max}, on the surface of freely suspended spherical particle, such a unicellular micro-organism, is given as:

$$\tau_{max} = 3\mu\gamma \tag{41}$$

When in suspension however cells tend to move at the minimum relative velocity with respect to the surrounding fluid. This means that a spherical cell will rotate at the velocity gradient of the suspension, or in proportion to it. For a spherical cell of diameter, d, the average shear rate, γ_{avg}, on its surface is given by:

$$\frac{u_s}{d} = \frac{\gamma_{avg}}{2} \tag{42}$$

where u_s is the surface velocity, and the average shear stress is:

$$\tau_{avg} = \frac{1}{2}\mu\gamma_{avg} \tag{43}$$

Moreover, a neighbouring cell will have a velocity equal but opposite in direction, giving a total surface velocity difference across the gap, b, between the two cells of $\Delta u_s = \gamma_{avg} d$. The average interstitial fluid velocity gradient is proportional to:

$$\left[\frac{d}{b}\right]\gamma_{avg} = \lambda\gamma_{avg} \tag{44}$$

where

$$\lambda = \frac{1}{\left[\frac{C_{max}}{C_v}\right]^{1/3} - 1} \tag{45}$$

in which C_v is the volumetric concentration and C_{max} ($= 0.74$ for monosized cells) is the theoretical maximum concentration (for $b = 0$). For monosized spherical cells suspended in a fluid of dynamic viscosity, μ, the resulting cell-fluid viscous interaction shear stress is given as [103]:

$$\tau_{p-f} = 2.25\,\mu\lambda^{\frac{3}{2}}\,\lambda_{avg} \tag{46}$$

The predictions in Fig. 20 shows that for low cell concentrations, the contribution to the total shearing stresses resulting from the particles fluid viscous interaction, Eq. (46), is negligible. However, as cell concentration increases, its contribution becomes increasingly more significant. As far as the authors are aware the effects of viscous fluid-cell interaction stresses on the disruption of biological materials has not been reported in the past. There are however process situations where this contribution may be important, for example in the regions of centrifuges and other separation units where relatively high solid concentrations exist.

Experimental data from capillary flow and rotating devices often demonstrate time-dependent disruption (see Fig. 12). Taking pipe flow as an example, assuming the flow to be fully developed, assuming a single pass and a fixed velocity through the capillary, the radial velocity gradient is often such that a micro-organism trapped in the flow will experience the greatest shearing stresses in a small zone close to the wall. The suspended micro-organisms will undergo random motion that will sometimes take them through the zone of the greatest shearing stresses. Once in that zone and based on the discussion on cell wall anelasticity, those micro-organisms that happen to be pre-aligned in such a way that the maximum stresses are imparted in the direction of the weakest orientation will stretch and break first. However, a large number of particles are likely to pass through the flow field without experiencing the highest shearing stresses. Of those that do only some are likely to stretch sufficiently to break. Against this background, consider what happens if the flow velocity through the pipe is maintained, but the number of passes through it is increased. The probability of exposure to the zone of highest shearing stresses will increase and so

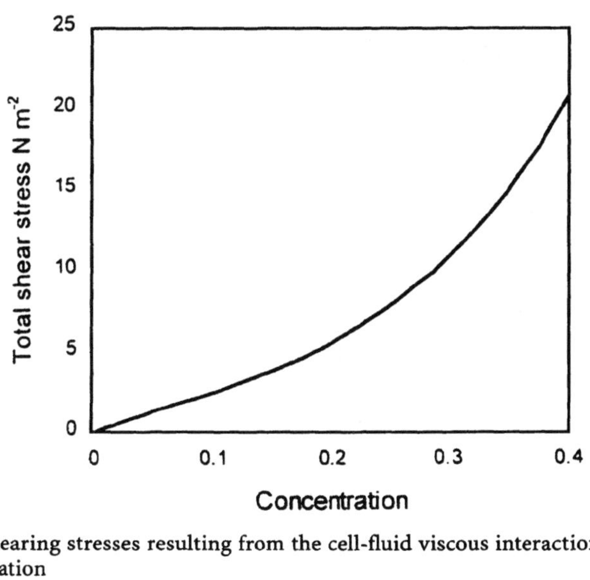

Fig. 20. Total shearing stresses resulting from the cell-fluid viscous interaction as a function of cell concentration

will the fraction of micro-organisms disrupted; a conclusion which is supported by ample experimental evidence [8, 16, 33, 104].

Limited experiments involving DNA breakage in a well-defined hyperbolic flow field produced by a 4-roll device have shown that laminar elongation stresses can also induce significant damage [30, 35]. Considerable work has been published on the breakage of freely dispersed drops in both shear and elongation flow. The general information that emerges from these studies provides a useful background to discuss the behaviour of bioparticles materials in similar flow situations.

The effects of elongation forces on a suspended drop are different to that of shear. For freely dispersed drops, it is universally recognised that the effects of shear and elongation on breakage depend critically on the ratio of the dispersed phase to continuous phase viscosity as shown in Fig. 21. It is well known that above a critical viscosity ratio of about 3.5 shear forces alone can not break drops where as elongational stresses are capable of breaking drops at both low and high viscosity ratios. In the context of micro-organisms with viscoelastic wall properties, it is the interfacial viscosity, μ, which is the appropriate parameter to be used in the definition of the viscosity ratio. Recalling the reported value of 10^{-3} Pa s for the interfacial viscosity of mammalian (red blood) cells and noting that the continuous phase has water-like properties ($\mu_c = 10^{-3}$ Pa s) gives a viscosity ratio of unity for red blood cells. According to the data in Fig. 21 the effect of elongation stresses on such cells are expected to be profound.

In a steady rotational shear field a spherical micro-organism with a soft wall will be deformed to an ellipsoidal shape. The deformation, M, is defined in

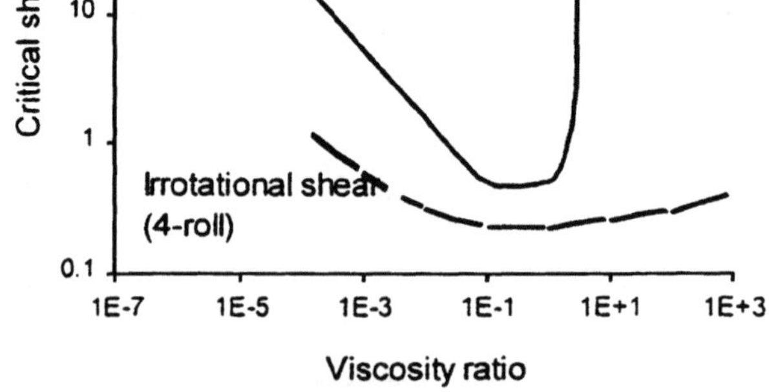

Fig. 21. Comparison of effect of viscosity ratio on critical shear in rotational (shear) and irrotational (extension) shear [76]

terms of the major dimensions of the deformed body (Fig. 22). Thus:

$$M = \frac{L - B}{L + B} \tag{47}$$

where L and B are the length and the breadth of the deformed micro-organism. According to Taylor [102] when the deformation is small, M is equal to the ratio of viscous to interfacial tension forces, the relationship between them is given by:

$$M = \frac{\gamma r \mu_{con} f_p}{\sigma_T} \tag{48}$$

where

$$f_p = \frac{19p + 16}{16p + 16} \tag{49}$$

where the viscosity ratio, p, is defined as:

$$p = \frac{\mu_{dis}}{\mu_{con}} \tag{50}$$

and μ_{dis} and μ_{con} are the dispersed and continuous phase viscosities respectively. Breakage occurs when deformation reaches a critical value, M_c. According to Rumscheidt and Mason [105, 109], for viscosity ratios in the range 0.1 – 1.0, and under certain conditions, breakage occurs when $M_c \approx 0.5 - 0.6$. For viscosity ratios lower than this range, M_c increases while for very high viscosity

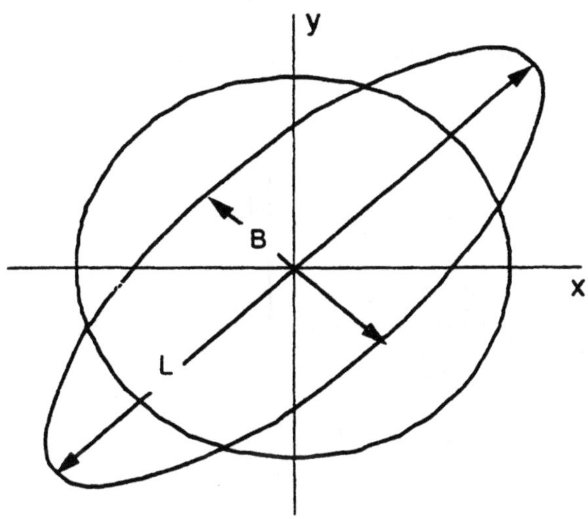

Fig. 22. Spherical drop deformed into an ellipsoidal shape in a rotational shear field [76]

ratios, in simple shear flow, deformation occurs at a limiting value of $M \approx 0.8\,p$ without breakage.

In contrast to rotational shear flow, deformation and breakage occurs over the whole range of viscosity ratio in an irrotational (extension) flow produced, for example, in a 4-roll apparatus (Fig. 23) from which the data shown in Fig. 21 were obtained [76]. Comparing the critical conditions for breakage by shear and by elongation, Fig. 23 shows that for equal deformation rates, irrotational flow tends to be more damaging than rotational flow.

Another important observation is that breakage occurs at the critical condition only if the time of imposed stresses acting on the drop, t_p, exceed the response time of the drop, t_r; the ratio of the two time scales is given by the drops time constant $d\mu/\lambda l$. If the imposed stresses are removed such that t_p is less than t_r, the drop will recover unbroken.

Figure 24 shows the dimensionless time defined as the ratio of t_p/t_r at the critical deformation for breakage plotted against the viscosity ratio for both irrotational and rotational flow fields. The plot indicates that for a fixed viscosity ratio the time required for breakage at the critical deformation is approximately the same for both shear and elongation. However, since according to the plots in Fig. 24 breakage by elongation occurs at lower critical deformations compared with shear it follows that for otherwise similar conditions the energy consumption for breakage due to elongation is less than that needed for shear [76].

Models based on Eqs. (47)–(50) have been used in the past to describe the disruption of unicellular micro-organisms and mammalian (hybridoma) cells [62]. The extent of cell disruption was measured in terms of loss of cell viability and was found to be dependent on both the level of stress (deformation) and the time of exposure (Fig. 25). All of the experiments were carried out in a cone and plate viscometer under laminar flow conditions by adding dextran to the solution. A critical condition for the rupture of the walls was defined in terms of shear deformation given by Eq. (44). Using micromanipulation techniques data were provided for the critical forces necessary to burst the cells (see Fig. 4)

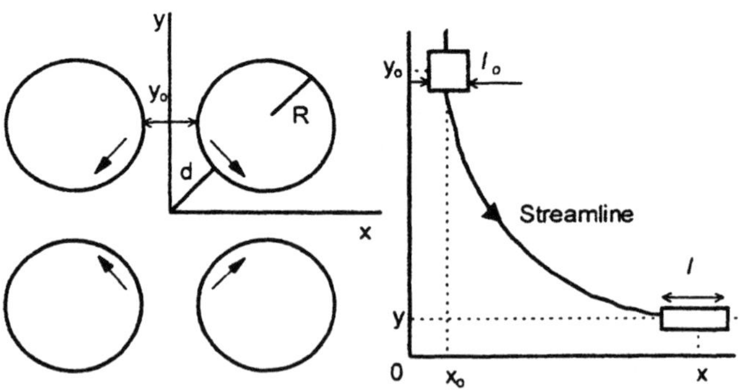

Fig. 23. Schematic diagram of the 4-roll mill apparatus. Schematic representation of the flow field within the mill illustrating the deformation of a fluid element [35]

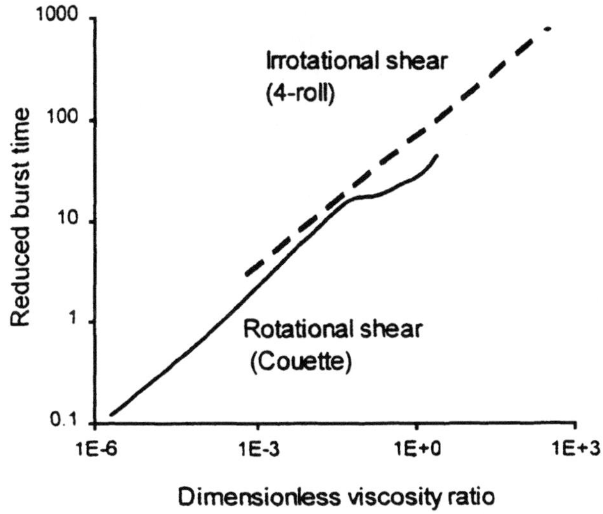

Fig. 24. Comparison of effect of viscosity ratio on reduced burst time for rotational and irrotational shear fields [76]

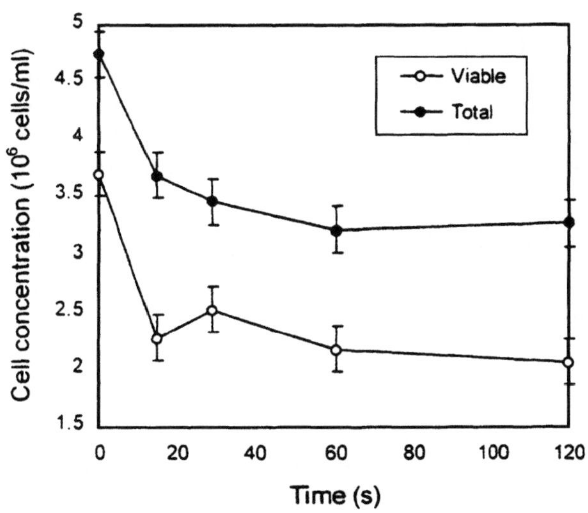

Fig. 25. Total and viable cell concentrations of TB/C3 hybridomas versus duration of shear in a cone and plate viscometer (shear stress 208 $N m^{-2}$). The error bars indicate the 95% confidence intervals [62]

and the results were found to be in good agreement with the model predictions. Measurements of solution viscosity were carried out routinely during the experiments but values were not quoted in the paper. It is therefore not possible to calculate the viscosity ratio, p (dispersed phase viscosity/continuous phase viscosity) for the system used by Born et al. [62]. Estimates of the solution viscosity based on the maximum shear stress and shear rate quoted in the paper however suggest significant elongational deformations during the experiments.

Elongation flow fields are likely to be present whenever streamlines cease to be parallel and in particular converging flow typical of that at the entrance of, for example, a capillary flow device is known to develop strong stretching flow. The strength of elongational flow at the entrance of a capillary can be determined by one of several equations [106]. In the case of capillary flow, an estimate of the normal (tensile) stress at the entrance can be obtained from:

$$\tau_{ent} = \frac{\tau_n}{\tan\alpha} \tag{51}$$

where τ_n is the shear stress at the nominal shear rate:

$$\gamma_n = \frac{4Q}{\pi r^3} \tag{52}$$

and α is the approach half angle.

3.4 Flow Associated Interfacial Effects

There is considerable experimental evidence indicating loss of biological activity of macromolecules such as globular proteins and enzymes at gas-liquid [57], liquid-solid (Fig. 26) [107] and liquid-liquid [108] interfaces. The extent of inactivation has been shown to be strongly influenced by the prevailing flow field and by, many other factors including the presence and/or absence of additives and contaminants and the type of solid surfaces (Figs. 27 and 28) [107].

In solution, thermodynamic considerations generally favour the diffusion of macromolecules to interfaces. In systems containing low concentrations of the macromolecules adsorption at an interface is likely to be diffusion controlled and leads to the formation of a relatively loose and mobile film at the interface. Changes in free energy, enthalpy and entropy associated with the adsorbed molecules cause instability in the structure of the adsorbed macromolecules. Consequently, unfolding and conformational changes of the adsorbed macromolecules occur leading to loss of biological activity. For otherwise similar conditions the degree of structural instability, and hence loss of activity, is highly system specific (Figs. 26–28). With increasing concentration of the macromolecules, the film thickness at the interface grows and the loss of activity becomes increasingly more dependent on the rate at which the film is disrupted and fresh interface is generated for adsorption. Dunnill and co-workers [19, 20, 24, 40, 51, 110] carried out extensive experiments in which they measured the loss of activity of range of macromolecules in solution under laminar and turbulent flow of shear rates up to $10^6\ s^{-1}$ (Figs. 29 and 30). They concluded that in the presence of a shear field the interfacial film can be considered to be in a dynamic state of formation and rupture at a rate which is determined by the flow intensity. Biological activity falls at a rate dependent partly on the surface

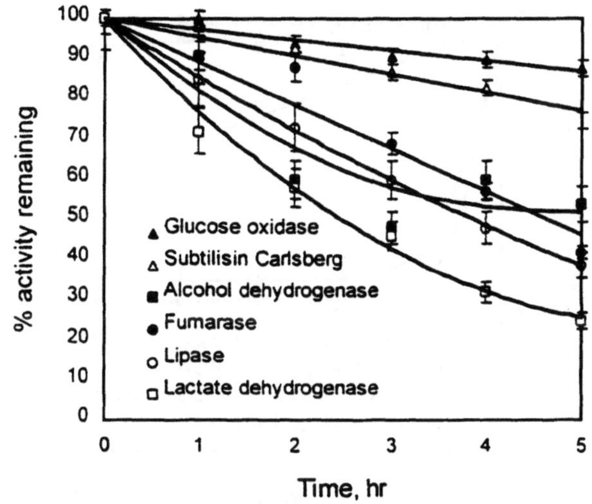

Fig. 26. The effect of shear on a range of enzymes with a stainless steel disc at a mean velocity gradient of 6490 s^{-1} at 30 °C, in the presence of an air/liquid interface. Each data point in the figure is the mean value of 4 replicates and is given as a percentage of the control value [107]

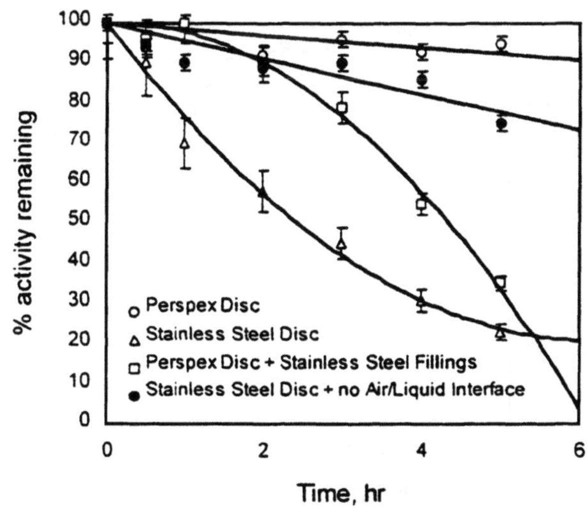

Fig. 27. The effects of shear on the inactivation rate of lactate dehydrogenase (LDH) at different conditions. LDH are subjected to shear at a mean velocity gradient of 6490 s^{-1} at 30 °C [107]

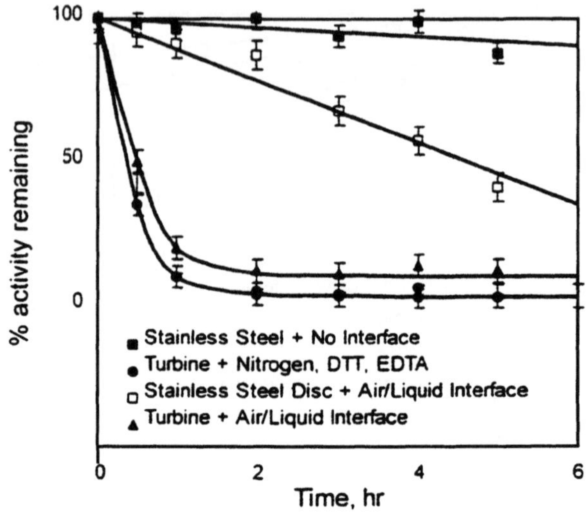

Fig. 28. Porcine Fumarase subject to shearing, with and without air/liquid interface, with a stainless steel disc at a mean velocity gradient of 6490 s^{-1} at 30 °C. The enzyme was also sheared, with and without air/liquid interface, with a 6 bladed Rushton turbine at a mean velocity gradient of 22700 s^{-1} at 30 °C [107]

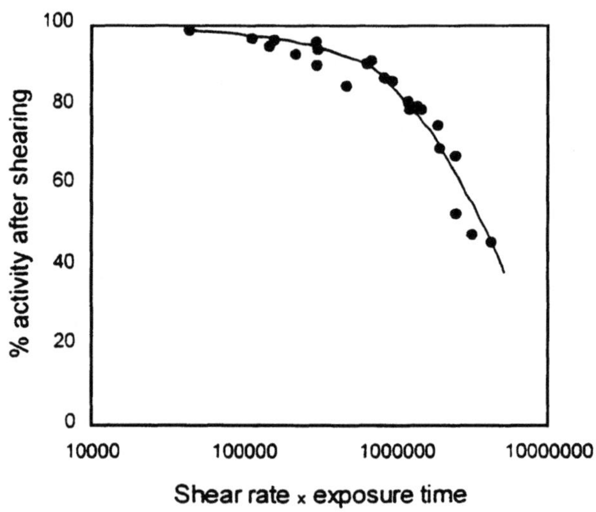

Fig. 29. Activity of catalase as a function of shear rate × time [42]

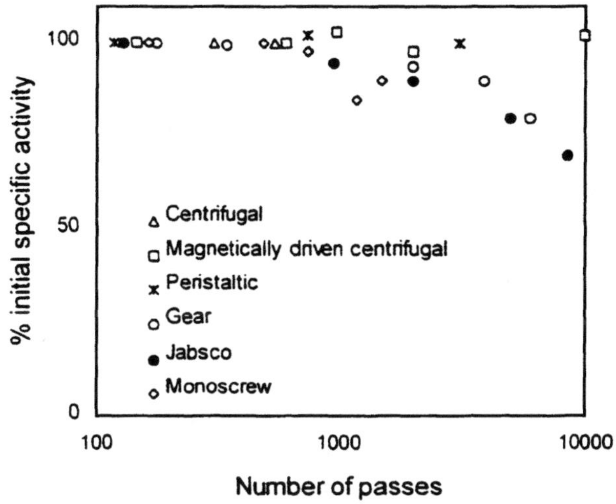

Fig. 30. Variation of alcohol dehydrogenase activity with number of passes through various pumps [42]

renewal rate, i.e. the shear rate, and partly by the physico-chemical properties and the dynamics of the unfolding of the macromolecules at the interface [110].

A number of researchers have argued that micro-organisms in suspension are also likely to suffer directly from shear stresses arising from flow of gas bubbles in process equipment. Estimates of typical flow stresses based on bubble formation, growth and detachment from spargers in airlifts and stirred vessels have been proposed and shown to be insignificant [112, 113]. The relative bubble-liquid velocity and the liquid motion in the wake behind a rising bubble have also been considered and flow stresses arising from these sources are thought to be relatively unimportant [114–116]. Numerical simulations have shown rapid bursting of bubbles at free surfaces to be one likely source capable of generating high local pressure stresses that can cause the disruption of micro-organisms that are in their neighbourhood [113, 117]. Figures 31 and 32 show numerical simulations of the maximum energy dissipation rate for bursting bubbles as a function of bubble radius. An important feature of this plot is the rapid exponential decay in the bubble burst energy dissipation rate as bubble size increases. Moreover the data in Figs. 31 and 32 represent the maximum energy dissipation rate. The average values of the energy dissipation rate are significantly smaller than the maximum, and a close examination of the energy dissipation profiles given in the original paper [118] indicates that the maximum values represent a very small region behind the bursting bubbles.

It is also worth noting that in an airlift or a mechanically stirred bioreactor small bubbles which correspond to a high energy dissipation rate during bursting are more likely to recirculate in the flow. In addition, in many biological systems of interest, gas bubbles are highly coalescing and thus bubbles that reach the free surface tend to be relatively large.

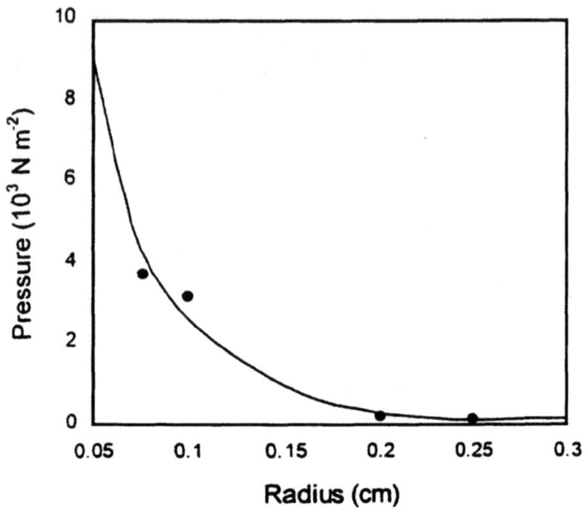

Fig. 31. Maximum pressures produced throughout the bursting process plotted against bubble radius [118]

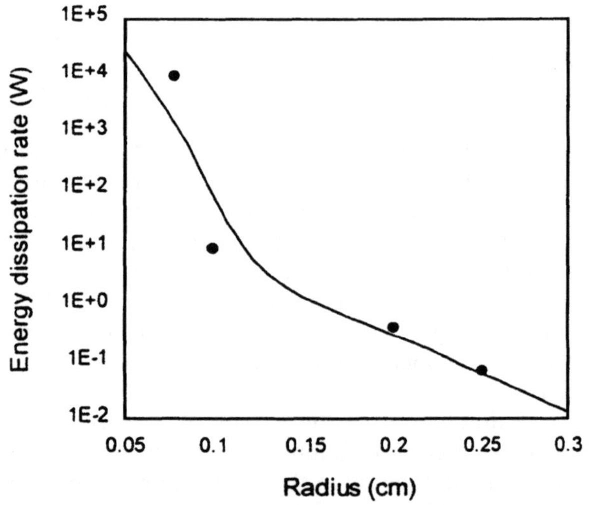

Fig. 32. Maximum energy dissipation rates produced throughout the bursting process, plotted against bubble radius. The logarithmic scale indicates an exponential dependence of maximum stress on bubble radius for large bubbles. The slight drop in the data point for the smallest bubble as compared to the next smallest may be because of the difficulty in locating the exact place and time of the peak, due to large spatial and temporal gradients beneath the forming jet [113]

4
Conclusions

Most biological materials are structurally weak and are thus easily damaged during processing, transportation and storage. Efficient process operations of "fragile" biomaterials must consider all the possible causes of damage. A review of the literature to date provides unequivocal evidence of both physical and biological damage to a broad range of biomaterials, from macromolecules in solutions to unicellular and filamentous micro-organisms and precipitates in suspension. Experimental and theoretical work published in the literature relates the main physical causes to shear and elongation flow stresses. In the vast majority of bioprocess operations, e.g. centrifugation, ultrafiltration, pumping and homogenisation, the engineering flow field in the equipment is often turbulent and the flow geometries are such that damage by shear and elongation flow pathways is not readily identifiable through normal analytical consideration of the flow field. Numerical techniques based on computational fluid dynamics (CFD) provides a means of simulating the flow field in complex geometries. Using such techniques it ought to be possible to identify the primary flow type and the intensity and duration of the stresses that are likely to be imposed on a process biomaterial during its passage through the equipment. Additionally, once identified, the primary flow type can be closely mimicked in purpose built small scale devices in which breakage pathways can be studied using small amounts of the biologically valued products under realistic flow conditions. Such an approach is currently being pursued in our laboratory.

It is generally recognised that the mechanical integrity of the walls of biomaterials is a key factor in determining both the likelihood and the kinetics of damage under a given set of conditions. A range of ill-defined qualitative descriptions is used including referring to the mechanical characteristics of the wall material. There is, however, a distinct lack of quantitative engineering information in the literature on the relevant biomechanical and rheomechanical properties of micro-organisms. In this literature review, we have attempted to draw upon classic information on the science of materials to identify the key biomechanical properties of micro-organisms in terms of an interfacial viscosity and a surface shear modulus of elasticity. It is the interaction between these material properties and the imposed flow stresses that determines the breakage pathways of a biomaterial in process operations. It is concluded that further progress in the optimisation of process operations requires quantitative data on the mechanical properties of such biomaterials. This is an area of immediate engineering interest and in need of further elucidation.

Finally, with macromolecules such as globular proteins, enzymes and DNA, degradation pathways have been identified based on gas-liquid, solid-liquid and liquid-liquid interfaces. Ample experimental evidence together with our understanding of the relevant thermodynamics supports the view that when in solution, biomacro-molecules favour interfaces where they tend to concentrate, unfold and undergo structural changes that can lead to loss of biological activity. The precise mechanisms of inactivation at interfaces are generally lacking, but there is strong experimental evidence indicating that the kinetics of interfacial inactivation critically depends on the intensity of the prevailing flow field.

Acknowledgements. The Advanced Centre for Biochemical Engineering is sponsored by the Biotechnology and Biological Sciences Research Council and the Council's support is gratefully acknowledged.

References

1. Abu-Reesh I, Kargi F (1989) J Biotech 9:167
2. Al-Rubeai M, Singh RP, Goldman MH, Emery AN (1995) Biotech Bioeng 45:463
3. Bhat VD, Windridge PA, Cherry RS, Mandel LJ (1995) Biotechnol Prog 11:596
4. Cherry RS (1993) Biotech Adv 2:279
5. Croughan MS, Hamel JF, Wang DIC (1987) Biotechnol Bioeng 29:130
6. Croughan MS, Sayre ES, Wang DIC (1989) Biotechnol Bioeng 33:862
7. Croughan MS, Wang DIC (1989) Biotechnol Bioeng 33:731
8. Dunlop EH, Namdev PK, Rosenberg MZ (1994) Chem Eng Sci 49:2263
9. Frangos JA, McIntire LV, Eskin SG (1988) Biotechnol Bioeng 32:1053
10. Garcia-Briones MA, Chalmers JJ (1994) Biotech Bioeng 44:1089
11. Goldblum S, Bae TK, Hink WF, Chalmers J (1990) Biotech Prog 6:383
12. Hooker BS, Lee JM, An G (1989) Enzyme Microbiol Technol 11:484
13. Hooker BS, Lee, JM, An G (1990) Biotechnol Bioeng 35:296
14. Leckie F, Scragg AH, Cliffe KR (1991) Enzyme Microbiol Technol 13:296
15. Lu GZ, Thompson BG, Gray, MR (1992) Biotech Bioeng 40:1277
16. McQueen A, Meilhoc E, Bailey JE (1987) Biotech Letters 9:831
17. Petersen JF, McIntire LV, Papoutsakis ET (1988) J Biotechnol 7:229
18. Maa YF, Hsu CC (1996) Biotech Bioeng 51:458
19. Thomas CR, Dunnill P (1979) Biotechnol Bioeng 21:2279
20. Thomas CR, Nienow AW, Dunnill P (1979) Biotechnol Bioeng 21:2263
21. Hagerman P (1988) An. Rev Biophys Biophys Chem 17:265
22. Kaya F, Heitmann JA, Joyce TW (1996) J Biotech 45:23
23. Virkar PD, Hoare M, Chan MYY, Dunnill P (1982) Biotechnol Bioeng 24:871
24. Virkar PD, Narendranathan NJ, Hoare M, Dunnill P (1981) Biotechnol Bioeng 23:425
25. Levinthal C, Davidson PF (1961) J Mol Biol 3:674
26. Theodossiou I, Collin IJ, Ward JM, Thomas, CR, Dunnill P (1997) Biopro Eng 16:175
27. Fishman DM, Patterson, GD (1996) Biopolymers 38:535
28. Carlson A, Signs M, Liermann L, Boor R, Jem KJ (1995) Biotechnol Bioeng 48:303
29. Marquet M, Horn NA, Meek JA (1995) BioPharm 9:26
30. Reese HR, Zimm BH (1990) J Chem Phys, 92:2650
31. Adam RE, Zimm BH (1977) Nucleic Acids Res 4:1513
32. Harrington RE, Zimm BH (1964) J Phy Chem 69:161
33. Bowman RD, Davidson N (1972) Biopolymers 2:2601
34. Davidson PF (1959) Proc Natl Acad Sci 45:1560
35. Odell JA, Taylor MA (1994) Biopolymers 34:1483
36. Harrison JS, Gill A, Hoare M (1998) Biotechnol BioEng (in press)
37. McCue JP, Sasagawa PW, Hein RH (1988) Biotechnol Appl Biochem 10:63
38. Young J, Dunnill P (1998) Bioprocess Eng (in press)
39. Ranjan V, Waterbury R, Xiao Z, Diamond SL (1996) Biotechnol Bioeng 49:383
40. Ayazi Shamlou P, Siddiqi SF, Titchener-Hooker NJ (1995) Chem Eng Sci 50:1383
41. Papoutsakis ET (1991) Trends in Biotech 9:427
42. Thomas CR (1990) Problems of shear in biotechnology. In: Winkler M (ed) Chemical engineering problems in biotechnology. Soc Chem Ind, Elsevier Applied Science
43. Whorlow RW (1980) Rheological Techniques. Ellis Horwood, Chichester

44. Hoare M (1982a) Trans IChemE 60:70
45. Hoare M (1982b) Trans IChemE 60:157
46. Hoare M, Narendranathan TJ, Flint JR, Heywood-Waddington D, Bell DJ, Dunnill P (1982) I EC Fundamental 21:402
47. Bell D J, Dunnill P (1982a) Biotechnol Bioeng 24:1271
48. Bell D J, Dunnill P (1982b) Biotechnol Bioeng 24:2319
49. Chan MYY, Hoare M, Dunnill P (1986) Biotechnol Bioeng 28:387
50. Glatz CE, Hoare M, Landa-Vertiz J (1986) AIChE J 32:1196
51. Bell, DJ, Hoare M, Dunnill P (1983) Adv Biochem Eng/Biotechnol 26:1
52. Twineham M, Hoare M, Bell DJ (1984) Chem Eng Sci 39:509
53. Devereux N, Hoare M, Dunnill P, Bell D J (1984) In: Gregory J (ed) Solid-Liquid separation. Chichester, Ellis Horwood Ltd, London
54. Devereux N, Hoare M, Dunnill P (1986a) Biotechnol Bioeng 28:88
55. Devereux N, Hoare M, Dunnill P (1986b) Chem Eng Commun 45:255
56. Bentham AC, Ireton MJ, Hoare M, Dunnill P (1987) Biotechnol Bioeng 31:984
57. Hoare M, Dunnill P (1989) Phil Trans R Soc Lond B324:497
58. Mannweiler K, Titchener-Hooker NJ, Hoare M (1989) IChemE Sym, Adv Biochem Eng, 105
59. Richardson P, Hoare M, Dunnill P (1989) Chem Eng Res Des 67:273
60. Davies PF, Remuzzi A, Gordon EJ, Dewey CF, Gimbrone MA (1986) Proc Natl Acad Sci 83:2114
61. Ayazi Shamlou P, Makagiansar HY, Ison AP, Lilly MD (1994) Chem Eng Sci 49: 2621–2631
62. Born C, Zhang Z, Al-Rubeai M, Thomas CR (1992) Biotechnol Bioeng 40:1004
63. Thorpe NO (1984) Cell Biology. John Wiley Sons, 3
64. Needham D, Ting-Beall HP, Tran-Son-Tay R (1991) Biotechnol Bioeng 38:838
65. Backman L (1986) J Cell Sci 80:281
66. Berk D, Evans E (1991) Biophys J 59:861
67. Bloom M, Evans E, Mouritsen OG (1991) Quart Reviews of Biophys 24:293
68. Evans EA (1983) Biophys J 43:27
69. Evans EA (1985) Biophys J 48:175
70. Fisher L (1993) J Chem Soc Faraday Trans 89:2567
71. Hochmuth RM, Wiles HC, Evans EA, McCown JT (1982) Biophys J 39:83
72. Waugh R, Evans EA (1979) Biophys J 26:115
73. Wessels JGH (1988) Acta Bot Neerl 37:3
74. Barnickel G, Naumann D, Bradaczek H, Labischinski H, Giesbrecht P (1983) The target of Penicillin. Walter de Gruyter, Berlin, 49
75. Phipps LW (1975) J Phys D: Appl Phys 8:448
76. Grace HP (1982) Chem Eng Commun 14:225
77. Namdev PK, Dunlop EH (1995) Applied BioChem Biotech 54:109
78. Ayazi Shamlou P, Stavrinides S, Titchener-Hooker NJ, Hoare M (1994) Chem Eng Sci 49:2647
79. Heydarian SM, Ison AP, Lilly MD, Ayazi Shamlou P (1998) Biotechnol Bioeng (submitted)
80. Olivier LA, Truskey GA (1993) Biotech Bioeng 42:963
81. Kolmogoroff AN (1949) Doklady Akad Nauk 66:825
82. Cherry RS, Papoutsakis ET (1986) Biopro Eng 1:29
83. Cherry RS, Papoutsakis ET (1988) Biotechnol Bioeng 32:1001
84. Cherry RS, Papoutsakis ET (1989) Biopro Eng 4:81
85. Ayazi Shamlou P (1993) Processing of solid-liquid suspensions. Butterworth Heinemann, London
86. Cherry RS, Kwon KY (1990) Biotechnol Bioeng 36:563
87. Aloi LE, Cherry RS (1994) J Biotech 33:21
88. Zhang Z, Al-Rubeai M, Thomas CR (1993) 42:987
89. Perry RH, Clinton CH (1973) Chemical Engineer's Handbook, McGraw-Hill, New York

90. Bird RB, Stewart WE, Lightfoot EN (1960) Transport Phenomena, Wiley
91. Davies JT (1972) Turbulence Phenomena, Academic Press, New York, international edition, John Wiley and Sons, Singapore
92. Edwards MF (1985) Mixing of low viscosity liquids in stirred tanks. In: Harnby N, Edwards MF, Nienow AW (eds) Mixing in the process industries. Butterworths, London
93. Gunkel AA, Weber ME. AIChE J 21:931
94. Placek J, Tavlarides LL (1985) AIChE J 31:1113
95. Schwartzberg HG, Treybal RE (1968) I EC Fundamentals 7:1
96. Cutter LA (1966) AIChE J 12:35
97. Wu H, Patterson GK (1989) Chem Eng Sci 44:2207
98. Kresta SM, Wood PE (1991) AIChE J 37:448
99. Aloi LE, Cherry RS (1996) Chem Eng Sci 51:1523
100. Zhou G, Kresta SM (1996) AIChE J 42:2476
101. Haas PA, (1987) AIChE J 33:987
102. Taylor GI (1934) Proc R Soc 146:501
103. Bagnold RA (1954) Proc R Soc Lond A225:49
104. Kieran PM, O'Donnell HJ, Malone DM, MacLoughlin PF (1995) Biotech Bioeng 45:415
105. Rumscheidt FD, Mason SG (1961) J Col Sci 16:210
106. Cogswell FN (1978) J Non-Newtonian Fluid Mech 4:23
107. Hoare M, Khan MR, Dunnill P (1992) Interfacial damage to proteins during intensive mixing in fermentation and down stream process. In: Van Den Tweei WJJ, Harder A, Buitelaar RM (eds) Stability and stabilization of enzymes, Proc Inter Sym, The Netherland, Elsevier Sci
108. Gargouri Y, Pieroni G, Riviere C, Sarda L, Verger R (1986) Biochem 25:1733
109. Rumscheidt FD, Mason SG (1961) J Col Sci 16:238
110. Hoare M, Narendranathan TJ, Flint JR, Heywood-Waddington D, Bell DJ, Dunnill P (1982) I & ED Fundamentals 21:402
111. Sadana A (1993), Bioseparation 3:297–320
112. Hua J, Erickson LE, Yiin TY, Glasgow LA (1993) Critical Rev Biotehnol 13:305
113. Cherry RS, Hulle CT (1992) Biotechnol Prog 8:11
114. Tramper J, Joustra D, Vlak, JM (1987) Bioreactor design for growth of shear-sensitive insect cell. In: Webb C, Mavituna R (eds) Plant animal cell cultures: process possibilities. Ellis Horwood, England
115. Glasgow LA, Jones GT, Erickson LE (1989) Hydrodynamic characterization of airlift bioreactor operation. In: Two-Phase Flows, Tapie, Tawan
116. Glasgow LA, Hua J, Yiin TY, Erickson LE (1992) Experimental studies of interfacial phenomena in sparged reactors. In: Tatterson GB, Calabrease RV (eds) Process mixing: chemical biochemical applications. AIChE Syposium Series 286
117. Garcia-Briones MA, Brodkey RS, Chalmers JJ (1994) Chem Eng Sci 49:2301
118. Boulton-Stone JM, Blake JR (1993) J Fluid Mech 254:437

Received August 1999

Influence of Stress on Adherent Cells

Gerlinde Kretzmer

Institut für Technische Chemie der Universität Hannover, Callinstraße 3,
D-30167 Hannover, Germany
E-mail: Kretzmer@mbox.iftc.uni-hannover.de

Stress is a broad term often used with animal cells. Frequently mechanical forces are meant using this term but chemical stress is also important cultivating animal cells. The chemical environment of the cell in a reactor have to be considered very carefully. The complexity of the medium requirements and the metabolic pathway cause very often growth limitations. Studying these limitations in order to find the reasons showed to be difficulty because of the complexity of the system. Nevertheless, glucose, glutamine, lactate and ammonia are found to be critical parameter as well as the osmotic pressure.

The influence of mechanical forces on cell viability is of great importance when growing the cells in agitated systems. By far the greatest amount of work reported in the literature has been done on suspension cells but adherent cells also experience shear forces not only in bioreactors also in vivo. Therefore, most research has be done on endothelial cells but studies exists done on non-endothelial cells. The influence of shear forces on cell growth, morphology and productivity will be discussed as well as possibilities of making the cells more resistant.

Keywords. Adherent cells, Chemical stress, Shear stress, Growth inhibition, Stress protection

Advances in Biochemical Engineering/
Biotechnology, Vol. 67
Managing Editor: Th. Scheper

Abbreviations

BHK	Baby Hamster Kidney
CHO	Chinese Hamster Ovary
FS-4	human diploid Fibroblast
MDCK	Madin Darby Canine Kidney
3T3	Murine Fibroblast
ATIII	Antithrombin III
b	width of the flow channel
d_{hy}	characteristic length of the channel
β-Gal	β-Galactosidase
h	height of the flow channel
L	Komolgorov length scale
L_e	entrance length of the channel
mRNA	messenger ribonucleic acid
Re	Reynolds number
t-PA	tissue plasminogen activator
UDP	Uridine diphosphate
UDP-GNAc	sum of UDP-*N*-acetylglucosamine and UDP-*N*-acetylgalactosamine
v	flow rate
ε	power dissipation per unit mass
η	dynamic viscosity of the medium
ν	kinematic viscosity of the medium
ϱ	density of the fluid
τ	shear stress

1 Introduction

Animal cells are exposed to various types of force regardless of whether growing in vitro or in vivo. The main and most severe forces are the mechanical forces nevertheless chemical forces cannot be neglected. Eukaryotes, such as animal cells, are multicellular, consisting of a variety of cell types specialized for different tasks. They need a quick exchange of substances and information between each other and therefore cannot be protected by a cell wall like bacteria. Animal cells have a plasma membrane consisting of a phospholipid bilayer. This makes the cell flexible and the interchange easier but on the other hand they lack the protection of the rigid cell wall.

Cultivating animal cells for production purposes is commonly known to be difficult, time consuming and expensive. This is caused by the complexity of the culture medium but also by the problems of reactor scaling up.

The chemical environment of the cells has to be considered very carefully. Animal cells do not need only a carbon and nitrogen source. They are dependent on a variety of other compounds (amino acids, vitamins, growth factors etc.).This complexity of requirements of the culture medium and the complexity of the metabolism hamper not only the development of defined media but also the recognition of inhibition and limitation effects. Animal cell are often exposed to chemical stress due to overfeeding or accumulation of metabolic end-products influencing cell growth and productivity.

Mechanical forces cause no problems as long as the cells grow under static conditions. Therefore, great efforts were made to scale up cultivation systems working under static conditions. Another reason for using static systems was the anchorage dependence of the most animal cells. Griffiths [1] gives a good overview of the possibilities. The development by Van Wezel [2] using microcarriers for growing adherent cells was epoch-making for it enables a real scale-up but the problem of mechanical forces is still present.

Scale-up to the industrial scale is only achievable and economical reasonable in stirred or airlift systems. To achieve high cell densities optimal nutrient supply is necessary. Therefore transport limitations have to be avoided by good mixing of cells (microcarrier) and medium. This movement caused by stirrer or aeration leads to mechanical forces evoking severe cell damage or cell death [3, 4].

To avoid this damage great efforts were made to improve the cultivation systems. Feder and Tolbert [5] developed special stirrers with low shear damage and Lehmann et al. [6] replaced gas bubbling through indirect aeration by diffusion of oxygen through porous polypropylene membranes.

Hollow fiber reactors [7] and dialysis reactors [8] avoid shear stress by separating cells and flowing media. In both reactors nutrient supply takes place by diffusion through the capillary wall or the dialysis membrane.

The main disadvantage of all these systems is the limitation of scale-up. Monoclonal antibodies are produced by multiplying the hollow fiber systems and stirred tank reactors with membrane aeration are known up to 100 liter. Small quantities of product can be produced by these systems but they are not suitable for real industrial scale-up.

Studying the forces causing these exposures will improve the optimization of production processes.

Not only for production purposes, the characterization of mechanical forces is of great interest. Medical research shows great interest in this field. In arteriosclerosis the influence of streaming blood on aortic cells is of interest [9]. This disease leads to an increase in the blood flow rate because of narrowing of the veins. In-vivo, shear-stress levels are in the range of 10 to 20 Nm^{-2}. Cells exposed to high shear stress showed alterations in their morphology and viability. Furthermore the effect of blood flow on smooth muscle cells after cardiovascular intervention is of great importance for the healing process [10].

2 Chemical Stress

The metabolism of animal cells is rather complex and this leads to complex growth requirements compared to bacterial cells. Therefore, cells were grown in undefined biological fluids for a long time. Eagle [11] was the first to reduce the number of defined compounds to those shown to be essential for cell growth.

Only two decades ago researchers started to study the effect of nutrient concentration on cell growth and metabolism systematically and the effect of produced metabolites. It appeared to be dependent on cell line, culture conditions and reactor [12]. It was found that the main carbon sources glucose and glutamine, and the metabolites lactate and ammonia as well as osmotic pressure could have negative effects on cell growth.

By far the greatest amount of work in the literature has been done on suspended hybridoma cells but the general tendency can be transferred to anchorage-dependent cell lines.

2.1 Nutrients

Glucose is the essential carbon source for most cell lines. It is normally added to the culture medium in a concentration of 5 to 25 mM. Small amounts of glucose are needed to provide ribose for the nucleic acid synthesis ($<25\ \mu M$), if the glucose concentration is increased the cell specific consumption rate increases dramatically [13]. Under these conditions, most glucose is converted into lactate following the anaerobic pathway [14]. This limited aerobic metabolism with high glucose concentrations is known as the Crabtree effect. Glucose itself does not inhibit cell growth but the fast conversion into lactate which is a toxic metabolite makes the glucose concentration into an important parameter.

Glutamine as an important source for the synthesis of purines, pyrimidines and amino sugars is essential for most cell lines, too. High concentrations of glutamine may also effect cell growth indirectly as the major end-products are lactate and ammonia [15]. Both known to be toxic metabolites.

2.2 Metabolites

Lactate is known to be the most important metabolic compound to effect cells negatively. It is released into the culture supernatant as a result of glycolysis and from glutamine conversion. In the literature, the effect of lactate concentration on cell growth is controversially reported. Hassel [16] found no effect on cell growth for a number of adherent cell lines if exposed to lactate concentrations higher than normally produced in batch cultures, whereas Wentz [17, 18] found a critical level for cultured BHK cells in T-flasks. Cells exposed to lactate concentration up to 2 $g\ l^{-1}$ did not show a negative response, increasing the level up to 2.5 $g\ l^{-1}$ reduced cell proliferation and a concentration higher than 2.5 $g\ l^{-1}$ inhibited cell growth. The morphology of cells exposed to higher lactate concentrations changed, cells undergoing division became larger than cells not exposed. Similar experiments in a bioreactor showed the same critical level but the effect was less dramatic [18]. These results make it evident that the lactate concentration is not the only parameter influencing cell growth and viability. Lactic acid production lowers the pH of the culture medium and this was found to be more important for hybridoma cells [19]. In a bioreactor the pH as well as the oxygen level is controlled.

Another toxic metabolite is ammonia. It is released by spontaneous decomposition of glutamine and to a far greater extend produced as a metabolite in the glutamine pathway. Concentrations of 2 mM and more as produced by adherent cells and hybridoma were reported to be growth inhibitory although the extend of inhibition varied between cell lines [16, 20, 21]. If the ammonia itself causes the inhibition of growth is questionable. Increasing ammonia concentrations increase the pH value of the culture medium. McQueen et al. [22] and Doyle et al. [23] showed that the changes in the pH with the absence of ammonia caused the inhibition of cell growth to the same extent as increasing the ammonia levels. Again it is to assume that changes in the pH value of the medium have a greater influence than ammonia itself. McQueen et al. [22] could not find a change in specific antibody production rate with decreasing pH and increasing ammonia levels whereas Miller et al. [24] and Glacken et al. [25] found a decrease in the specific productivity with increasing ammonia. Ryll et al. [26] studied the nucleotide pool in dependence on ammonia concentration. They found a correlation of the intracellular amount of UDP-GNAc (sum of UDP-*N*-acetylglucosamine and UDP-*N*-galactosamine) with the level of growth inhibition. The increase of UDP-GNAc is strictly dependent on the presence of glucose and ammonia. Glucose deficient media do not show such an increase. The extent of the increase in the UDP-GNAc level varied with the cell line. Cell lines adapted to ammonia learn to grow with higher UDP-GNAc levels.

McQueen et al. [22] could not find an influence of ammonia or pH on product quality, whereas Papoutsakis et al. [27] found significant alterations in expression rates and glycosylation pattern of recombinant mouse placental lactogen proteins produced in adherent CHO. They suggest a careful consideration of culture pH and ammonia concentration to ensure proper and reproducible quality of glycosylated proteins. Ryll et al. [26] also suggest that the change in the intracellular UDP-GNAc pool may affect the glycosylation pattern of recombinant proteins since the UDP-GNAc precursor D-glucosamine is known to show such an affect.

2.3
Osmotic Pressure

Controlling the pH value of the culture is often done by addition of sodium chloride and this changes the osmotic pressure of the culture supernatant. Such an alteration also occurs in fed-batch cultures where single compounds are added during cultivation. Omasa et al [28] showed that lactate did not have the inhibitory effect on hybridoma growth rate but rather the changes in osmotic pressure caused by adjusting the pH value. They adjusted the pH with sodium chloride and sodium lactate, respectively and found the same influence in both experiments. Oh et al. [29] showed that hyperosmotic conditions suppressed hybridoma growth severely but if adapted to grow in high osmolarity media the cells increase their productivity. Wentz [17] showed that osmolarity of 495 mosM and higher had a negative effect on the adherence of BHK cells. The morphology of the cell and the membrane properties changed, the cell shrank.

Most work on chemical stress was done with hybridoma cells since they are widely used for industrial relevant monoclonal antibody production but the overall tendency is also true for anchorage-dependent cells. This was shown in the few papers published but is also well known to researchers in laboratory.

3
Shear Forces

The main reasons for the damage to cells in a reactor are the apparent shear forces and the collision of microcarriers with themselves and with turbulent eddies. In the literature studies are mainly focused on suspension cells and there again on hybridoma cells. The work reported in the literature can be divided into two fields: studies dealing with the influence of various stirrer speeds on cell viability and those investigating the influence of defined shear forces on cells with a viscosimeter.

The shear forces are mainly in the range of 1 to 10 Nm^{-2}. This exposure causes cell death between 20 and 80% depending on the exposure duration which is between a few seconds and several hours. Studies performed in a bioreactor have an exposure duration of several days. The results are partly contradictory. Tramper et al. [30] found a critical stress level of 1.5 Nm^{-2} for insect cells, whereas Oh et al. [31] could not show an influence on hybridoma cells even at high stirrer speed. This shows that each cell line reacts different and that there is a necessity for defined stress systems if the results is to be comparable.

Concerning adherent cells there are few studies in the literature. Some of them deal with the influence of stirrer speed on microcarrier cultures. Most studies using defined forces are from medical research. These studies, as well as those with production cells, use different types of exposure systems based on the parallel plate theory. They investigate the influence of stress on cell morphology and viability which is most important for arteriosclerosis research.

Table 1 gives an overview on the main literature.

3.1
Concept of Method

Calculation of the shear forces cells would experience in a reactor during cultivation is very complex. In order to initiate process development under the proper conditions, it is essential to know the physical conditions under which the cells are grown. Aunins et al. [32] characterized the effect of impeller geometry, impeller height and agitation speed on power input and mass transfer in a 500-ml spinner vessel. The power input data show significant deviation from literature correlation. The power input is much lower than predicted and this reflects the milder turbulence in the small-scale vessel. The data point out differences in small vessel physics that may give odd results when investigating hydrodynamic effects in cell culture.

Two different concepts are described in the literature for investigating the influence of shear stress on adherent cells: Microcarrier cultures in stirred reactors and defined stress levels in flow chambers.

Table 1. Summary of the main literature on the influence of shear forces on adherent cells

Cell type	Test apparatus	Exposure level	References
Kidney cells	Flow chamber	0.2–6 Nm^{-2} (2–24 h)	[44]
Endothelial cells	Flow chamber	1 Nm^{-2} (7–8 h)	[55]
Endothelial cells	Flow chamber	1–8.5 Nm^{-2} (0–24 h)	[9]
Endothelial cells	Flow chamber	1–3 Nm^{-2} (0–24 h)	[45]
Endothelial cells	Flow chamber	0–2.4 Nm^{-2} (7–8 h)	[56]
Fibroblast	Microcarrier	60–220 rpm (0–150 h)	[36]
Fibroblast	Microcarrier	Microcarrier concentration	[37, 38]
BHK	Flow chamber	0–80 Nm^{-2} (0–3 h)	[47]
Endothelial cells	Flow chamber	0.2–2.5 Nm^{-2} (0–30 h)	[57]
BHK	Flow chamber	0–2.5 Nm^{-2} (24 h) critical level	[48]
BHK	Flow chamber	0–4 Nm^{-2} (24 h) cell division	[49]
BHK	Flow chamber	0–4 Nm^{-2} (24 h) temperature	[50]
BHK	Flow chamber	0–4 Nm^{-2} (24 h) cholesterol addition	[51]
rBHK	Flow chamber	0–4 Nm^{-2} (24 h) productivity	[52]
Endothelial cells	Coverslips & Microcarrier	150 rpm	[41]
Endothelial cells and tissue	Flow chamber		[54]
MDCK cells	Coverslips	0.002–0.027 Nm^{-2}	[42]
Endothelial cells	Flow chamber	3 Nm^{-2} (24 h)	[53]
Human aortic smooth muscle cells	Flow chamber	0.5–2.5 Nm^{-2} (24 h)	[10]
Human umbilical vein cells	Flow chamber	0–1 Nm^{-2}	[46]

3.1.1 *Microcarrier Cultures*

Cherry and Papoutsakis [33] refer to the exposure to the collision between microcarriers and influence of turbulent eddies. Three different flow regions were defined: bulk turbulent flow, bulk laminar flow and boundary-layer flow. They postulate the primary mechanism coming from direct interactions between microcarriers and turbulent eddies. Microcarriers are small beads of several hundred micrometers diameter. Eddies of the size of the microcarrier or smaller may cause high shear stresses on the cells. The size of the smallest eddies can be estimated by the Kolmogorov length scale L, as given by

$$L = (\nu^3/\varepsilon)^{1/4} \tag{1}$$

where ν is the kinematic viscosity and ε is the power dissipation per unit mass.

The ratio of the Komolgorov length scale and the diameter of the microcarriers correlate with cell damage. Several researchers show that if the ratio falls below one, cell damage increases significantly [34–38]. The size of the smallest eddies increases with an increase in kinematic viscosity and a decrease with power input. As the size of the eddies should be larger than the diameter of the microcarrier this gives two possibilities for avoiding damage: addition of substances to increase the viscosity or a decrease in agitation rate. Croughan et al. [38] showed the beneficial effect of thickening agents added to the culture medium. They found that a three-fold increase in viscosity leads to a 10-fold decrease in cell damage of human diploid fibroblasts (FS-4) caused by hydrodynamic forces.

Cell damage can also be dependent on the concentration of microcarriers. Croughan et al. [37] found that high concentrations of microcarriers cause greater damage at high agitation speed, whereas Hu et al. [39] found that they do not at low agitation speed.

Investigating the kind of damage the cells experienced, Croughan et al. [40] postulated that agitation killed the cells, whereas Cherry et al. [35] found a reduction in growth rate. These results obtained with different cell lines show again the problematic situation comparing results from different studies and postulating a general concept of the influence of shear stress on cells.

Cherry [41] and Bhat et al. [42] reported on physiological studies on 3T3 cells on microcarriers. They studied 3T3 cells grown on coverslips under static and under agitation conditions (150 rpm) and compared the results with 3T3 cells grown on microcarriers. The highest growth rate was found under static conditions and the lowest under agitation conditions on coverslips. Surprisingly, the cells grown on microcarriers showed an intermediate growth rate. Stress fibers in the actin cytoskeleton were found in the cells grown on converslips whereas they were lacking in cells grown on microcarriers. The number of focal contacts which a cell adhere to its substratum and the mitochondria followed the same pattern of reduced numbers when grown on microcarriers. Since Cherry [41] compared cells grown on coverslips under agitation and cells on microcarriers, it is obvious that the stirrer speed itself is not the only reason for the alterations. Cells growing on freely suspended microcarriers experience different conditions than cells grown on plane static surface even when the medium is agitated.

3.1.2
Flow Chamber Systems

By far the greatest amount of work reported in the literature has been done on adherent cells exposed in flow chambers. This concept offers two advantages: For experiments in biomedical research on endothelial and smooth muscle cells the conditions in a flow chamber resemble the most the conditions in vivo and for adherent producer cells the flow chamber gives the possibility to expose them to defined shear stress. In agitated cultures in a reactor, a variety of forces occur causing viability problems. In addition, each cell will experience a different set of forces which contribute to the status of the cell. For investigating the influence of shear stress on physiological parameter, defined conditions are essential, concomitant flow chambers can be constructed to monitor the cells during exposure.

Flow chambers are based on the theory of parallel plates. They should provide a defined two-dimensional laminar flow of medium over a monolayer of cells. Based on this theory Levesque et al.[9] described an equation for the calculation of the shear stress. Shear stress τ is then given as

$$\tau = 6\, v\, \eta / h^2\, b \tag{2}$$

were v is the flow rate, η is the dynamic viscosity of the fluid, h is the height of the channel and b is the width of the channel.

In order to calculate the shear stress in the chamber several parameter have to be fulfilled: To achieve a two-dimensional flow the ratio of width to height of the flow channel is 5:1 [43]. To maintain a laminar flow the Reynolds number, given as

$$Re = \frac{\nu\, \varrho\, d_{hy}}{\eta} \tag{3}$$

should be less than 2000 (were ϱ is the density of the fluid and d_{hy} is the characteristic length of the system given by $d_{hy} = 4$ (area/perimeter).

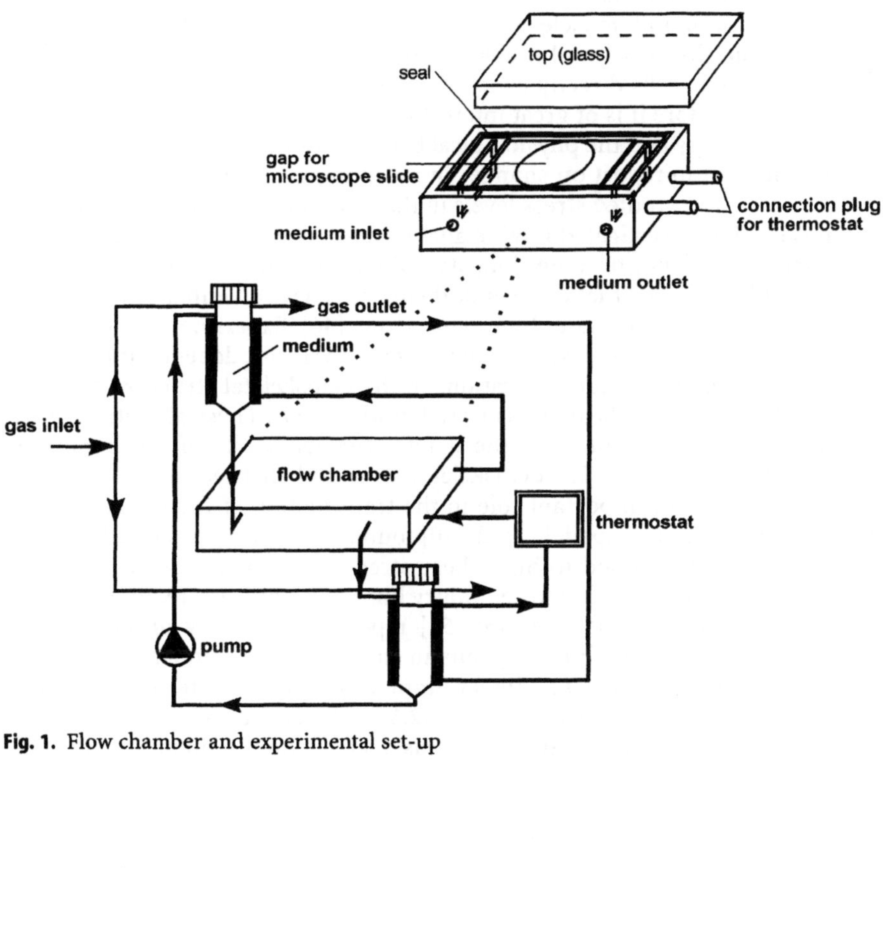

Fig. 1. Flow chamber and experimental set-up

To maintain a laminar flow profile over a long distance the entrance length (L_e) of the flow channel has to be small compared to the total length. The entrance length is given by

$$L_e = 0.05\ h\ Re \tag{4}$$

After this length the laminar profile is stable.

Several flow chambers are described in the literature [9, 44–47]. They are all based on the above mentioned theory. Figure 1 shows the improved flow chamber and the experimental set-up we used in our experiments [48–52].

The flow chamber consists of a lower part and an upper part. A depression was milled into the stainless steel lower part and the inlet and outlet for the medium and supplies for temperate water were connected to it. The cells grow on glass slides which were fitted into the depression. The upper part was made of glass in order to provide visual control of the cell monolayer. The flow chamber was arranged between two medium reservoirs at different heights to achieve a laminar flow. The medium circulated continuously and the oxygen and carbon dioxide level and temperature were maintained. With this set-up, it was possible to monitor the cells and their movements using a film camera outside the incubator. Yosikawa et al. [47] also described a parallel-plate chamber using a confocal laser scanning microscope to monitor the cells and their calcium-ion content.

Flow chambers are mainly used in biomedical research for studying endothelial cells and smooth muscle cells. The arterial endothelium located between blood flow and the vessel wall is always exposed to shear stress in vivo as well as smooth muscle cells directly exposed to blood flow after cardiovascular intervention. Therefore it is of great interest to study the influence of shear stress on the morphology and the physiological behavior of these cells. Levesque et al. [9] were the first to report on changes in morphology of the cells on exposure to this stress. They studied stress levels in the range of 1–8.5 Nm^{-2} for 0 to 24 h of exposure. The investigated bovine aortic endothelial cells oriented with the flow direction and they become elongated when exposed to higher shear stress. Further studies showed that these endothelial cells have an enhanced uptake of low-density lipoproteins under high shear stress [46]. Experiments with thymidine indicate that the increased uptake observed was not dependent on an increase in cellular replication. Alterations in the cytoskeletal structure of endothelial cells exposed to shear stress were found [53, 54]. These alterations were referred to as rearrangement in the F-actin filaments, one of the important cytoskeleton components. The cytoskeleton of endothelial cells and focal adhesion proteins play an important role in the transduction of forces causing the elongation of the cells. Papadaki et al. [10] found a reduction in smooth muscle cell number while exposed to high shear stress (2.5 Nm^{-2}). They could show that the decrease was not due to cell injury but to a significant reduction in the proliferation rate. Frangos et al. [55, 56] reported very early on changes in prostacyclin production of primary human endothelial cells under shear stress. The production rate increased linear from 20 ng per minute to 120 ng per minute with increasing shear stress (0–2.5 Nm^{-2}) and exposing the cells to pulsatile shear stress the enhancement was again 2 times higher. Nollert et al.

[57] found an increase in mRNA-production of the investigated t-PA when exposing human endothelial cells for 24 h with 1.5 and 2.2 Nm^{-2}, respectively. Whereas the mRNA production of the determined endothelinpetide decreased at the same time. These results prove that there is a dependence between shear stress, productivity and transcription.

It is questionable if these results obtained with endothelial cells can be transferred to other non-endothelial cells. Because of their origin endothelial cell are structured to withstand hydrodynamical stress.

Investigations with non-endothelial cells carried out in flow-chambers showed different reactions on shear stress. The critical level for cell detachment during short time exposure of Baby Hamster Kidney (BHK), Vero and MRC-5 cells was found to be in the range of 3–5 Nm^{-2} [58]. Stathopoulos [45] reported an elongation of human embryonic kidney cells after exposure to 0.65 Nm^{-2} for 24 h. They also found an increased release of urokinase under these conditions. BHK cells change their morphology under shear stress as well. They were shortened under higher stress until they were triangular and finally rounded up [44]. Shear stress levels up to 1 Nm^{-2} over a period of 24 h show no effect on BHK morphology, whereas increasing the stress level BHK cells show the above mentioned alterations in shape and an orientation of the cell perpendicular to the flow [48]. This behavior in shape and orientation was proved by a time lapse film (not published). Cells were torn away from the substratum under high shear stress (<2.5 Nm^{-2}) destroying their membranes. Moderate shear stress leads to alterations as reported above. The perpendicular orientation of the cells occurred by loosening one end of the cell and shortening towards the other end. The cells performed this in and against flow direction. The influence of shear stress on mitosis could also be studied with the mentioned time lapse film. The time for division of the mother cell was not influenced by exposure to shear stress, whereas the spreading time after mitosis was increased. Exposure level of 4.5 Nm^{-2} resulted in a 3-times longer spreading time [49]. Up to 2 Nm^{-2} there was no significant loss of cells during mitosis, whereas at a level of 4.5 Nm^{-2} nearly 30% of the mitotic cells were lost by detachment.

With a set of parameters studied during shear stress experiments a critical shear stress level for recombinant and non-recombinant BHK cells could be determined [48, 52]. Figure 2 shows the critical level by calculating the LDH-Release Rate as a marker for cell death.

Table 2 shows the critical levels for 3 different BHK cell lines for the most important parameter.

Table 2. Summary of the parameters indicating the critical shear stress level (Nm^{-2})

	BHK 21	rBHK (β-Gal)	rBHK (ATIII)
Cell Viability	0.75–1.0	≅ 0.7	< 0.6
Morphology	< 1.0	n.d.	n.d.
Cell Size Distribution	0.8–1.0	n.d.	0.4–0.7
LDH-Activity	0.75–1.0	≅ 0.7	≅ 0.5
Productivity	n.d.	≅ 0.7	0.6–0.7

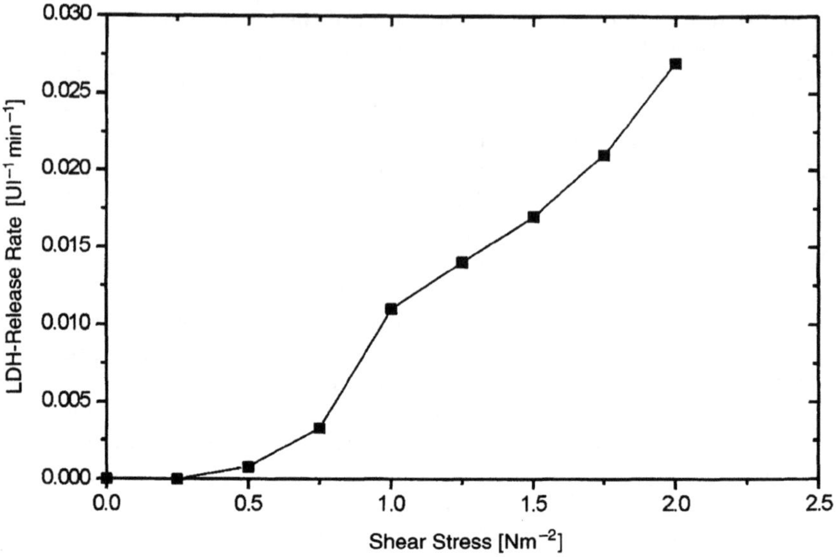

Fig. 2. LDH-Release Rate as a function of shear stress indicating the critical level exposing adherent BHK cells for 24 h

The effect of overall shear stress on adherent cells is not necessarily negative. The results suggest that independently from the genetic structure the critical shear stress level is in the range of 0.7 to 1.0 Nm^{-2}. Moderate shear stress is beneficial for proliferation and productivity. Exposing cells to shear stress levels less than 1 Nm^{-2}, cell growth is enhanced compared to cultures under static conditions. Cell specific productivity also rose with moderate shear stress. Productivity of BHK-(β-Gal) cells was 2 times higher and of BHK-(ATIII) 1.5 times higher than under static conditions [59].

A lot of effort is made to avoid shear stress in bioreactors. Looking at the results for endothelial and non-endothelial cells, moderate shear stress is desirable in order to enhance proliferation and specially productivity. Therefore research should focus on methods of how to make cells withstand shear stress rather than how to avoid shear stress in a reactor.

In principle, there are two methods to achieve this goal: Adding shear stress protective agents to the culture medium or changing the physical properties of the cells. Addition of polymers like polyvinyl alcohol, polyethylene glycol, or pluronic [60–62] in order to lower the negative influence of shear stress as it is done with suspension cells is not reasonable for adherent cells. The detergent properties of these agents will spoil the adherence of the cells to their substratum.

We were able show that temperature is a convenient and promising parameter to achieve this goal. Growing cells at temperatures lower than 37 °C will slow down growth rate but on the other hand have a beneficial effect on shear resistance [50] (Fig. 3).

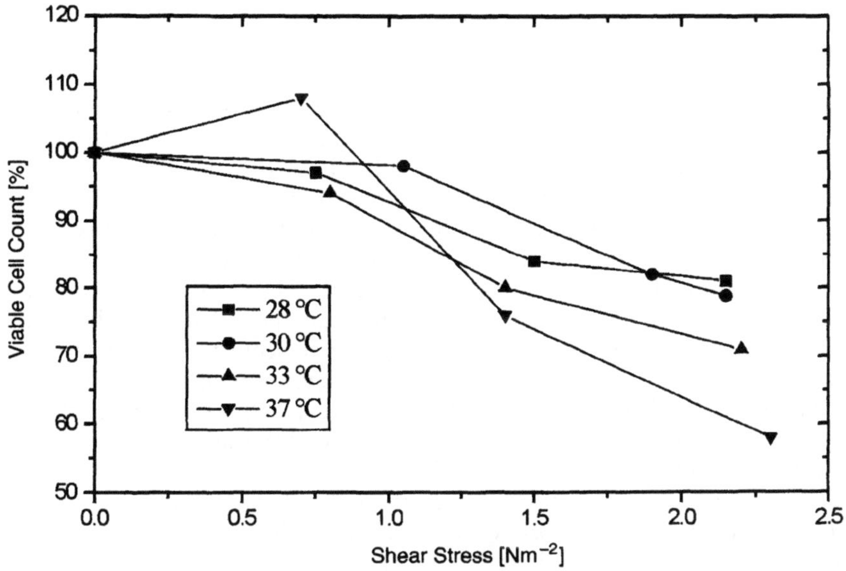

Fig. 3. Viable cell count of adherent BHK cells as a function of temperature and shear stress

The critical level was only slightly shifted growing the cells at 30 °C but the negative effect on cell viability was less significant than at 37 °C. Cell viability at a shear stress level of about 2.2 Nm^{-2} and an exposure duration of 24 h dropped at 37 °C down to 58% whereas the viability at 30 °C was still 79%. A temperature of 28 °C showed best results but there was no cell growth visible. We suggest that the beneficial effect of temperature is connected to changes in the cell membrane composition. Temperature is known to alter the membrane fluidity. Lowering the temperature, decreases the fluidity of the membrane and makes the cell more rigid. This, we assume, helps the cell to have a better shear stress resistance. To prove this assumption we studied the influence of cholesterol addition to the culture medium [51]. Cholesterol is the ingredient of the membrane responsible for the fluidity. Increasing cholesterol concentrations (30 µg ml^{-1} up to 217 µg ml^{-1}) in the culture medium showed similar effects on cell growth to lowering temperature. With cholesterol addition there was a beneficial effect on cells exposed to shear stress. Up to 90 µgml^{-1} cell viability was 15% higher after exposure to 1.5 Nm^{-2} for 24 h compared with cultures under static conditions.

This proves our theory that the membrane fluidity is an important parameter increasing shear stress resistance. The studies give two possibilities of improving the resistance: lowering temperature or adding cholesterol. Which one is the most convenient is dependent on the cell line and the constraints of the process.

4 Conclusion

Animal cells experience a variety of stresses regardless of being grown in vivo or in vitro. The influence of the chemical environment on cell growth, viability and productivity as well as on product quality is hard to determine because of the complexity of the system. Sometimes the inhibition effect is caused indirectly as it is for glucose and glutamine. However, if the obviously toxic metabolites lactate and ammonia cause the actual damage is questionable. Some researchers assumed and tried to prove that effects these substances have on the chemical environment are more important than the substances themselves. Changes in culture pH and the osmotic pressure seem to be more harmful. Research in this area is still going on and the development of more and more defined media with fewer ingredients will be very helpful in solving the problems and understand the biochemical influence of the substances.

In the field of mechanical stress on adherent cells there is still a great demand especially for producing cells. In the field of medicine, research is still going on and becoming more systematic. Researchers have investigated not only the influence on growth and viability but also on morphology and productivity. For producing cells, the amount of data is still small and fragmentary. It is questionable if the data obtained for endothelial cells can be transferred to the non-endothelial cells. There are only few data on cell viability and morphology and even fewer on the influence of shear stress on productivity. Some results show that moderate shear stress is beneficial for production processes but research on possibilities for increasing the shear stress resistance of the cells is still at its beginning.

The overall conclusion is that there is already some data available on the influence of stress on adherent cells but there is still important data missing to understand the behavior of adherent cells in bioreactors.

References

1. Griffiths B (1992) Scaling-up of animal cell cultures. In: Freshney RI (ed) Animal cell culture A practical approach. Oxford University Press, Oxford, p 47
2. Van Wezel AL (1967) Nature 216:64
3. Handa-Corrigan A, Emery AN, Spier RE (1989) Enzym Microb Technol 11:230
4. Papoutsakis ET (1991) TIBTECH 9:427
5. Feder J, Tolbert WR (1983) Scientific American 248:36
6. Lehmann J, Piehl GW, Schulz R (1987) Dev Biol Standard 66:227
7. Jäger V (1988) Dissertation Universität Hannover
8. Märkl H, Lechner M, Götz F (1990) J Ferment Bioeng 69:244
9. Levesque MJ, Nerem RM (1985) J of Biomechan Eng 107:341
10. Papadaki M, McIntire LV, Eskin SG (1996)Biotechnol Bioeng 50:555
11. Eagle H (1955) Science 122:501
12. Lambert KJ, Birch JR (1985) In: Spier RE, Griffiths JB (eds) Animal Cell Biotechnology. Academic Press, London 1:86
13. Zielke HR, Ozand PT, Tildon JT, Sevdalian DA, Cornblath M (1978) J Cell Physiol 95:41
14. Glacken MW, Fleischaker RJ, Sinskey AJ (1988) Biotechnol Bioeng 28:1376

15. Zielke HR, Sumbilla CM, Sevdalian DA, Hawkins RL, Ozand PT (1980) J Cell Physiol 104:433
16. Hassell T (1988) PhD thesis, Manchester Polytechnic, UK
17. Wentz D (1989) PhD thesis, University of Hannover, Germany
18. Wentz D, Schügerl K (1992) Enzyme Microbiol Technol 14:68
19. Reuveny S, Velez D, Macmillan JD, Miller L (1986) J Immunol Methods 86:53
20. Butler M, Spier RE (1984) J Biotechnol 1:187
21. Butler M (1985) Dev Biol Stand 60:269
22. McQueen A, Bailey JE (1990) Biotechnol Bioeng 35:1067
23. Doyle C, Butler M (1990) J Biotechnol 15:91
24. Miller WM, Blanch HW, Wilke CR (1988) Biotechnol Bioeng 32:947
25. Glacken MW, Adema E, Sinskey AJ (1988) Biotechnol Bioeng 32:491
26. Ryll T, Valley U, Wagner R (1994) Biotechnol Bioeng 44:184
27. Papoutsakis ET, Linzer DIH, Borys MC (1994) In: Animal Cell Technology: Products of Today, Prospects for Tomorrow, Butterworth-Heinemann, Oxford, 658
28. Omasa T, Higashiyama K, Shioya S, Suga K (1992) Biotechnol Bioeng 39:556
29. Oh SKW, Vig P, Chua F, Teo WK, Yap MGS (1993) Biotechnol Bioeng 42:601
30. Tramper J, Williams JB, Joustra D, Vlak JM (1986) Enzyme Microbiol Technol 8:33
31. Oh SKW, Nienow AW, Al-Rubeai M, Emery AN (1989) J of Biotechnol 12:45
32. Aunins JG, Woodson BA, Hale TK, Wang DIC (1989) Biotechnol Bioeng 34:1127
33. Cherry RS, Papoutsakis ET (1986) Bioproc Eng 1:29
34. Cherry RS, Papoutsakis ET (1988) Biotechnol Bioeng 32:1001
35. Cherry RS, Papoutsakis ET (1989) Bioproc Eng 4:81
36. Croughan MS, Hamel JF, Wang DIC (1987) Biotechnol Bioeng 29:130
37. Croughan MS, Hamel JF, Wang DIC (1988) Biotechnol Bioeng 32:975
38. Croughan MS, Sayre ES, Wang DIC (1989) Biotechnol Bioeng 33:862
39. Hu WS, Meier J, Wang DIC (1985) Biotechnol Bioeng 27:585
40. Croughan MS, Wang DIC (1989) Biotechnol Bioeng 33:731
41. Cherry RS (1993) Biotech Adv 11:279
42. Bhat VD, Windridge PA, Cherry RS, Mandel LJ (1995) Biotechnol Prog 11:596
43. Krüger JW, Young DF, Cholvin NR (1971) J of Biomech 4:31
44. Stathopoulos NA, Hellums JD (1985) Biotechnol Bioeng 27:1021
45. Sprague EA, Steinbach BL, Nerem RM, Schwartz C (1987) Circulation 76:648
46. Yoshikawa N, Ariyoshi H, Ikeda M, Sakon M, Kawasaki T, Monden M (1997) Cell Calcium 22:189
47. Kretzmer G, Schügerl K (1991) Appl Microbiol Biotechnol 34:613
48. Ludwig A, Kretzmer G, Schügerl K (1992) Enzyme Microbiol Technol 14:209
49. Ludwig A, Tomeczkowski J, Kretzmer G (1992) Biotechnol Lett 14:881
50. Ludwig A, Tomeczkowski J, Kretzmer G (1992) Appl Microbio Biotechnol 38:323
51. Tomeczkowski J, Ludwig A, Kretzmer G (1993) Enzyme Microb Technol 15:849
52. Ludwig A, Kretzmer G (1993) J of Biotechnol 27:217
53. Thoumine O, Ziegler T, Girard PR, Nerem RM (1995) Exp Cell Research 219:427
54. Sato M, Ohshima N (1994) Biorheology 31:143
55. Frangos JA, Eskin S, McIntire LV, Ives CL (1985) Nature 227:1477
56. Frangos JA, Eskin S, McIntire LV (1988) Biotechnol Bioeng 32:1053
57. Nollert MU, Diamond SL, McIntire LV (1991) Biotechnol Bioeng 38:588
58. Spier RE, Crouch CE, Fowler H (1985) Dev Biol Standard 66:255
59. Ludwig A (1993) PhD thesis, University of Hannover, Germany
60. Michaels JD, Petersen JF, McIntire LV, Papoutsakis ET (1991) Biotechnol Bioeng 38:169
61. Al-Rubeai M, Emery AN, Chalder S (1992) Appl Microbiol Biotechnol 37:44
62. Smith CG, Greenfield PF (1992) Biotechnol Bioeng 40:1045

Received September 1999

[illegible]

Effects of Hydrodynamic and Interfacial Forces on Plant Cell Suspension Systems

Patricia M. Kieran[1] · Dermot M. Malone[2] · P. Frank MacLoughlin[3]

Department of Chemical Engineering, University College Dublin, Belfield, Dublin 4, Ireland.
[1] *e-mail: patricia.kieran@ucd.ie*
[2] *e-mail: dermot.malone@ucd.ie*
[3] *e-mail: frank.macloughlin@ucd.ie*

Plant cells are perceived to be sensitive to the hydrodynamic environment in conventional bioreactors. Heightened sensitivity, relative to most bacterial cultures, is frequently attributed to larger plant cell sizes, extensive vacuolization and aggregation patterns. Early studies of shear sensitivity focused on cell lysis and/or loss of viability. More recently, an extensive array of sub-lethal responses has been identified. A fuller understanding of these sub-lytic effects may assist in the optimization of large-scale cultivation conditions. This paper reviews recent work on the hydrodynamic shear sensitivity of plant cell systems, under cultivation conditions and in purpose-built shearing devices. The relevance of different approaches to the characterization of the intensity of a given hydrodynamic environment is discussed. Indicators of cell response to hydrodynamic stress are evaluated. The potential significance of cellular defense mechanisms, observed in response to mechanical stimulants, is identified.

Keywords. Plant cell suspensions, Hydrodynamic shear, Energy dissipation, Aeration, Oxidative burst

Advances in Biochemical Engineering/
Biotechnology, Vol. 67
Managing Editor: Th. Scheper

List of Symbols and Abbreviations

ALR	air-lift reactor
AOS	active oxygen species
C	constant in Eq. (13)
C_1	constant in Eq. (15)
d	diameter (m)
D	impeller diameter (m)
E	cumulative energy ($J\,kg^{-1}$)
ECP	extracellular polysaccharide
HR	hypersensitive response
HRGP	hydroxyproline-rich glycoproteins
k	first order rate constant
K	constant in Eq. (12)
$k_L a$	mass transfer coefficient (h^{-1})
l	length (m)
LDH	lactate dehydrogenase
N	impeller rotational speed (s^{-1})
N_P	power number
OB	oxidative burst
OUR	oxygen uptake rate
P	power (W)
p	pressure ($N\,m^{-2}$)
PGA	polygalacturonic acid
Q	volumetric flow rate ($m^3\,s^{-1}$)
r	radius (m)
Re	Reynolds number
STR	stirred tank reactor
t	time
TTC	2,3,5-triphenyltetrazolium chloride
u	velocity ($m\,s^{-1}$)
V	volume (m^3)
vvm	volumes per volume per minute
wv	working volume
x	axial distance from nozzle (m)
X	viability
y	dimensionless radial distance (r/x)
ε	energy dissipation rate ($W\,kg^{-1}$)
γ	shear rate (s^{-1})
θ	exposure time (s)

μ	dynamic viscosity ($Ns\,m^{-2}$)
ν	kinematic viscosity (m^2s^{-1})
ϱ	density ($kg\,m^{-3}$)
τ	shear stress ($N\,m^{-2}$)

Subscripts

0	at t = 0
c	critical
e	eddy
g	gas
k	Kolmogoroff
L	liquid
m	center line
o	outlet
p	particle
Re	Reynolds
w	wall

1 Introduction

Plant cell suspensions offer the potential to produce valuable phytochemicals, traditionally extracted from the naturally grown whole plant, under controlled and reproducible conditions. To date, commercial processes involving these systems have been limited to just a handful of applications, including the much-cited shikonin [1] and ginseng [2, 3].

However, the promise of plant cell culture is clearly indicated by the enormous range of other systems which have been investigated to date (e.g. [4–6]). The failure of suspension cultures to make a widespread industrial impact is essentially due to economic feasibility [5, 7] which, assuming that a viable market for a plant product exists, derives from a combination of biological and process-based factors. From an engineering perspective, the primary challenges are process design and process scale-up [8, 9]. Loss of productivity associated with scale-up from shake flask to laboratory or pilot-scale reactors has been commonly reported, although 75-m^3 systems have been successfully operated [10]. Recently, Curtis [11] has suggested that many of the difficulties encountered in the cultivation of plant cells, particularly in bioreactors originally designed for microbial systems, arose from a lack of understanding of the specific requirements of plant suspensions. One aspect of this work which has attracted considerable attention in recent years is the perceived sensitivity of plant cell suspensions to the hydrodynamic environment encountered in a bioreactor.

The susceptibility of biological systems, including procaryotic and eucaryotic cultures and enzyme solutions, to the forces prevailing under normal processing conditions has been extensively studied and is the subject of comprehensive reviews [12–16], including other chapters in this volume. Downstream processing operations, as well as routine pumping, will expose cell suspensions to significant shearing effects. However, most research effort has focused on the response of cellular systems to the forces prevailing under cultivation conditions in a bioreactor. Compared with the more commercially significant mammalian systems, comparatively little work has been performed on plant cell suspensions [17] and much of it is highly system-specific. However, recent reviews by Prokop and Bajpai [12] and Namdev and Dunlop [18] have taken a more integrated approach to the analysis of plant cell systems.

The purpose of this chapter is to review relevant work in the area, including an evaluation of methods employed for shear sensitivity studies, comparison of bases for data analysis and an outline of current knowledge of the interface between the hydrodynamic environment and the plant cells themselves. Coverage is limited to cell suspension cultures and does not extend to other tissue or organ systems.

2 Plant Cells in Suspension Culture

The perceived sensitivity of plant cells to the hydrodynamic stress associated with aeration and agitation conditions is typically attributed to the physical characteristics of the suspended cells, namely their size, the presence of a cell wall, the existence of a large vacuole, and their tendency to aggregate. Table 1 illustrates some of the differences between plant cells and other biological systems. Chalmers [19] attributed shear sensitivity in mammalian cultures at least in part to the fact that these cells occur naturally as part of a tissue, surrounded by other cells. The same is true for plant cells. The more robust microbial systems, on the other hand, exist in nature as single organisms or mycelial structures, very close to the forms they assume in submerged culture.

In suspension, plant cells are significantly larger than most microbial cells and are typically of the order of 10–100 μm in size. They vary in shape from cylindrical to spherical. The plasma membrane is surrounded by a primary cell wall which defines the cell size and shape. The robustness of plant cells, relative to mammalian cells or to plant protoplasts [18], is usually attributed to the pre-

Table 1. Comparison of biocatalyst characteristics [8]

Biocatalyst	Shape	Size (μm)	Cell wall	Aggregates	Doubling time (h)	OUR (mmol l^{-1} h^{-1})
Plant cells	Spherical; Cylindrical	l: 100–500; d: 20–50	Yes	Yes	20–100	10^1
Animal cells	Spherical	d: 10–20	No	No	20	10^1
Bacteria	Spherical; Cylindrical; Spiral	d: < 1; l: < 5	Yes	Yes/No	0.5–10	10^3
Yeasts	Spherical	d: 5–10	Yes	Yes/No	10	10^2
Moulds	Mycelial	d: 5–10; l: < 100	Yes	Yes/No	10–20	10^2

sence of this cell wall. The cell wall allows the cell to maintain turgor, and tensile strengths of the order of 10–30 atm have been reported by Carpita [20] for suspension cultured carrot cells. However, the cell wall is not merely a rigid capsule and its dynamic nature is reflected in its role in osmo-regulation and cell expansion [21]. The vacuole, which acts as a storage compartment but which is also involved in cell expansion and cell turgor [22], can occupy up to 50% of the cell volume and typically increases in size over the course of a batch growth cycle.

Plant cells in suspension culture seldom exist in isolation and aggregates, with diameters of many millimetres, may consist of hundreds of cells. Aggregate shape and size varies with cell line, cultivation conditions and growth stage. Aggregation is typically attributed to the failure of cells to separate after division, coupled with the secretion of extracellular polysaccharides (ECPs) which may help to bind cells. However, the presence of cytoplasmic channels in cell walls, known as plasmodesmata, suggests that intercellular communication may be an important process in system metabolism. Aggregation patterns, coupled with high biomass concentrations and ECP secretions, result in high viscosity broths, typically exhibiting non-Newtonian, shear-thinning, characteristics [23–31]. Although volumetric oxygen requirements are characteristically lower than microbial systems (Table 1), high broth viscosities tend to reduce oxygen mass transfer coefficients in bioreactors. Increases in agitation/aeration intensity intended to counteract this phenomenon may exacerbate what are collectively identified as "shear effects", as well as other phenomena including foaming and gas-stripping. A full elucidation of the interplay between the flow field in a bioreactor and plant cells or aggregates in suspension must take account of these distinguishing features.

3 Hydrodynamic Characterization of the Cultivation Environment

The stirred tank reactor (STR) is still the most commonly used reactor in plant cell suspension processes, although there are many reports in the literature of applications involving alternative bioreactor designs [32], including rotating-drum reactors [1, 33, 34], fluidized-bed reactors [35] and bubble columns and air-lift loop reactors (ALRs) [36, 37]. However, because of their ubiquity, STRs have been most frequently employed in the study of shear effects on plant cells. Characterization of the turbulent flow field in a mechanically agitated bioreactor is a complex problem. This topic has been treated in some detail in recent, comprehensive reviews of the shear sensitivity of biocatalysts in general [12, 14, 38], and of the hydrodynamics of STRs [39, 40] in particular. In this chapter, coverage is limited to an overview of approaches to characterizing the hydrodynamic environment with specific reference to plant cell suspension culture applications.

Classical analysis of turbulence is based on single-phase systems. The presence of suspended particles is known to alter the nature of the turbulent field, although the precise mechanism is, as yet, poorly defined [41]. For particles of average diameter d_p, the degree of modification is thought to be related to the ratio d_p/l_e, where l_e is the length of the energy-containing eddies. Effects are more severe for extreme phase density ratios (e.g. gas-solid or gas-liquid) and high solids loading than for the more moderate phase density ratios and solid volume fractions typically encountered in bioprocess applications. However, limited data are available for three-phase systems. Turbulence may also be affected by macromolecules present in the liquid, potentially either medium constituents or components secreted by the cells during cultivation [42]. In the analysis of turbulence phenomena in STRs, isotropic conditions are generally assumed; however, only in the region very close to the impeller is this likely to apply [40]. The assumption of a constant local value of energy dissipation rate also neglects the intermittent nature of turbulence found even in regions where isotropic turbulence prevails [38].

For hydrodynamic and specifically shear considerations, two distinct regions are usually identified in an STR: (a) the high shear region(s) surrounding the impeller(s), typically comprising between 1 and 5% of the total broth volume, in which local turbulent energy dissipation rates may be as high as 100 times the tank averaged rate $\bar{\varepsilon}$ [43], and (b) the remainder of the vessel, wherein the local energy dissipation rate may fall to 0.25 $\bar{\varepsilon}$ [44]. The average energy dissipation rate, $\bar{\varepsilon}$, is calculated as

$$\bar{\varepsilon} = \frac{P}{\rho V} = \frac{N_P N^3 D^5}{V} \tag{1}$$

where N is the rotational speed of the impeller (s^{-1}), D is the impeller diameter (m), N_P is the power number for the relevant impeller, at the prevailing value of the Reynolds number, and V is an appropriate dissipation volume.

For plant cell suspensions cultivated in shake flasks, Huang et al. [45] used the energy dissipation rate as a correlating parameter for system response. Specific power input was calculated using the empirical correlation proposed by Sumino et al. [46] and subsequently employed in other applications [47, 48]:

$$\frac{P}{V} = 1.09 \times 10^{-6}\, V_L^{-0.25}\, N^{2.81} \tag{2}$$

where the specific power input (P/V) is expressed in kWm^{-3}, V_L is the liquid volume in the flask (ml) and N is the shaker rotational speed (rpm). While the intensity of the hydrodynamic environment in the shaker flask will, of course, depend on flask geometry (e.g. whether baffled or unbaffled), Eq. (2) can provide a useful estimate of the relevant energy dissipation.

For the purposes of shear related studies, a variety of methods have been employed to characterize the hydrodynamic environment in a bioreactor. Many are based on the agitation rate, including impeller rotational speed (N), impeller tip speed (πND) and the "integrated shear factor" [$2\pi ND/(T-D)$] [49]. None are suitable for comparative purposes across different reactor geometries. The applicability of impeller tip speed is further limited by the fact that, irrespective of vessel scale, a comparatively narrow range of tip speeds is employed in conventional STRs. The preceding parameters consider only agitation conditions. To take cognisance of both aeration and agitation effects, a number of workers [27, 50–53] have employed the concept of an initial mass transfer coefficient as an empirical indicator of the intensity of the environment. In recent years, energy dissipation rates [45, 54–56] and, in particular, cumulative energy dissipation (the product of energy dissipation rate and exposure time) [57–59], have been identified as more appropriate measures of the intensity of a cell's experience in a given hydrodynamic environment. In an agitated bioreactor, a representative specific power input of 1 kWm^{-3} corresponds to daily cumulative energy dissipation levels of approximately 9×10^7 Jm^{-3}.

In the analysis of shear effects in turbulent flows, the concept of the Kolmogoroff length scale has received considerable attention. It is commonly assumed [60] that suspended particles will be harmlessly entrained by eddies in the so-called Kolmogoroff range which are considerably larger than the particle diameter, but will be susceptible to damage from encounters with eddies of the same order of magnitude or smaller. The characteristic length (l_k) and velocity (u_k) of these eddies are described by the following equations:

$$l_k = \left(\frac{\upsilon^3}{\varepsilon}\right)^{1/4} \tag{2}$$

$$u_k = (\upsilon\varepsilon)^{1/4} \tag{4}$$

Average energy dissipation rates in an STR are of the order of 1 Wkg^{-1}, although levels can vary by an order of magnitude between laboratory and industrial scale vessels. Local dissipation rates may be of the order of 100 Wkg^{-1}. For water the corresponding values of l_k lie between 10 and 30 μm. As broth viscosity changes over the course of the growth cycle, the Kolmogoroff eddy length will change. Assuming that isotropic turbulence applies, the portion of the aggregate size distribution susceptible to eddy-related damage will also vary. Single plant cells have typical diameters of 25–50 μm. Aggregates, however, have diameters of between 100 and 1000 μm, or greater. Although the Kolmogoroff length scale approach has been widely applied to animal cell systems, the dimensions of plant cell aggregates suggest that they may be more susceptible to eddies in the inertial subrange. However, the effects will still depend on the dominant aggregate size. Huang et al. [45] extended the model developed by Croughan et al. [61] for microcarrier cultures to suspensions of *Stizolobium hassjoo.* The kinetic equations to describe biomass production incorporated an aggregate disruption rate constant. Experimentally determined variations in average aggregate size were found to be quite well described by the proposed model, even allowing for the fact that the aggregate size distributions showed both considerable spread and evidence of bimodality. The aggregate disruption rate increased with energy dissipation rate (shaker speed) and fell with increasing broth viscosity over the course of the growth cycle. For STR-cultivated suspensions of *Morinda citrifolia*, at an agitator energy input of approximately 1.5 Wkg^{-1}, and assuming that all energy was dissipated in the impeller region, Kieran [29] calculated Kolmogoroff length scales of the order of 100 μm, which is similar in magnitude to individual cells, both in length (~50–100 μm) and diameter (~35 μm). However, chains consisting of 10 cells were typically an order of magnitude larger. Moreover, with l/d ratios in excess of 10, it is arguable that models derived for spherical particles are of limited practical relevance.

Hydrodynamic effects on suspended particles in an STR may be broadly categorized as time-averaged, time-dependent and collision-related. Time-averaged shear rates are most commonly considered. Maximum shear rates, and accordingly maximum stresses, are assumed to occur in the impeller region. Time-dependent effects, on the other hand, are attributable to turbulent velocity fluctuations. The relevant turbulent Reynolds stresses are frequently evaluated in terms of the characteristic size and velocity of the turbulent eddies and are generally found to predominate over viscous effects.

Considering time-averaged flow effects, the shear rate in the bulk fluid is frequently assumed to be directly proportional to the impeller speed [62–64]. Although this simple relationship does not apply to turbulent flow, it is commonly employed as a general indicator of the hydrodynamic environment. For transitional and turbulent flow, for fluids of specified rheology and a range of vessel geometries, expressions have been proposed for bulk average (e.g. [65]) and maximum [66–68] shear rates. On the basis of both bulk and maximum shear rates, Rosenberg [69] employed the analyses of Matsuo and Unno [70] and Schlichting [71] to estimate bulk and maximum (impeller boundary layer) stresses, respectively. Carrot cell aggregates of 190 μm diameter, suspended in a low viscosity Newtonian fluid, were sheared in a rotoviscometer, equipped with a turbine impeller. At tip speeds of between 0.1 and 4 $m s^{-1}$, the aggregates were estimated to be subjected to bulk shear stresses in the range 0.01–6.0 $N m^{-2}$; the corresponding boundary layer stresses were in the range 0.1–90 $N m^{-2}$. Using a broadly similar approach for 100-μm particles subjected to a tip speed of 3 $m s^{-1}$, in 20,000- and 4-l vessels, Meijer [72] predicted maximum stresses of between 18 and 40 $N m^{-2}$. For a 2-l STR, agitated at a tip speed of 1.4 $m s^{-1}$, Dunlop et al. [57] calculated bulk time-averaged shear stresses of 0.08 $N m^{-2}$.

For certain conditions, the hydrodynamic stresses generated by the very rapid fluctuations in turbulent flow, the so-called Reynolds stresses, can be estimated as [70]:

$$\tau_{Re} \cong \varrho (\varepsilon d_p)^{2/3} \qquad (5)$$

For the same bioreactor mentioned above (operating at an impeller tip speed of 1.4 $m s^{-1}$), Dunlop et al. [57] predicted maximum Reynolds stresses in the impeller region of 32.4 $N m^{-2}$. However, if the energy is assumed to be uniformly dissipated throughout the vessel contents, then Eq. (5) will yield lower values. As calculated, the Reynolds stresses involve a length scale and the stress experienced by a particular entity will depend on its size.

In addition to bulk liquid turbulence effects, suspended particles may be involved in collisions with one another or with solid surfaces within the vessel. This phenomenon has been extensively studied in micro-carrier cultures [60] and appears to be significant at high concentrations [61]. Rosenberg [69] and Meijer [72] applied the approach of Cherry and Papoutsakis [60] to the study of collision phenomena involving spherical plant cell aggregates of 190 and 100 μm, respectively. In both cases it was concluded that for typical biomass concentrations, particle–particle interactions were of less significance than particle–impeller collisions.

4
Response Indicators for Shear Effects

The response of a biological system to any imposed stress will depend on the physiological characteristics of the system, as well as on the nature, intensity and duration of the relevant stress conditions. While "stress" is most commonly perceived as detrimental, the system response may be either positive or negative. For example, it is well established that secondary metabolism in whole plants can be stimulated by pathogenic attack and this phenomenon has been profitably exploited in the elicitation of cell suspension systems [73, 74]. In considering hydrodynamic-related stress, however, emphasis has tended to focus on negative or damaging effects. In STRs, positive effects associated with an increase in the intensity of the hydrodynamic environment may often be attributed to an associated enhancement of mass and heat transfer rates. In plant cell studies, considerable emphasis has focused on critical or lethal effects. However, in terms of the optimization of plant cell suspensions, sub-lethal metabolic phenomena are extremely important. Moreover, identification and quantification of such responses may elucidate the mechanism by which hydrodynamic stress signals are received and transduced within the cell.

4.1
Loss of Viability

Because the ability of a cell to grow and divide is fundamental to the production of biomass and/or metabolites, loss of viability is the most commonly investigated response to hydrodynamic stress. The most rigorous and arguably most valuable method for establishing the viability of a sheared sample is via regrowth studies under control conditions, in shake flasks, for example [29, 55, 75]. A technique described by Horsch and Jones [76] has been employed [57] in regrowth studies of sheared carrot suspensions. However, due to slow growth rates and inherent susceptibility to contamination, regrowth studies are necessarily lengthy and inconvenient.

Cell lysis is the most extreme response of cells to an imposed hydrodynamic or interfacial stress. For suspensions investigated under short-term, non-growth conditions, the simplest method for assessing cell lysis is by microscopic determination of cell numbers before and after shearing. Aggregated suspensions may require maceration prior to counting in order to separate the cells [72]. Hooker et al. [77] evaluated the extent of cell lysis in *N. tabacum* suspensions on the basis of the reduction in dry weight concentration over the course of short-term shear trials; in fully lysed samples, there was a 75% reduction in biomass dry weight relative to unlysed cells. In actively growing cultures subjected to hydrodynamic stress, it is inherently more difficult to distinguish between variations in cell number attributable to lysis and those associated with stress-induced reductions in cell growth; microscopic observation of cellular debris [78] may be useful as a qualitative indicator of lysis.

4.2
Membrane Integrity

An alternative and indirect method of evaluating loss of viability is in terms of the integrity of the selectively permeable cytoplasmic cell membrane, most frequently assessed using staining techniques. For plant cell systems, the stains most commonly used are Evan's Blue [79], fluorescein diacetate and phenosafranin [80]. Choice of an appropriate stain may depend on the nature and extent of cell pigmentation. For example, studies in this laboratory have revealed that neither Evan's Blue or phenosafrinin is appropriate for use with suspensions of *Drosera capensis* as intracellular anthocyanin masks the stain. With fluorochromes, the time taken for fluorescence to develop may also be affected by pigmentation [81]. Percentage viability is determined by microscopically counting the number of stained/unstained cells. Large numbers of cells must be counted for statistical accuracy, and, unless automated, the method is subject to human counting errors. Tucker et al. [82] have reported the measurement of viability in hybridoma cells stained with Trypan Blue, using fully automated image analysis. For plant systems, color-based image analysis has been used [83] to identify pigmented cells in *Ajuga* suspensions. However, there are no reports of automated viability measurements.

Membrane integrity in plant cell suspension cultures has been measured using dielectric spectroscopy [84], which is based on the dielectric properties of the suspension. This technique offers the potential for on-line evaluation of cell viability, and data for a range of cell cultures showed good agreement with other indicators of system performance. However, the results may be affected by significant changes in cell size over the course of an investigation.

A dual isotope labeling technique [85] has been used to measure membrane permeability in plant cells, based on the selective permeability of the membranes of living cells to tritiated water and carbon-14 labeled mannitol. Kieran [29] showed that the results of the dual isotope labeling and Evan's Blue staining methods correlated well as indicators of cell viability; however, the latter was preferable in terms of reagent cost and ease of analysis.

It is important to note that all of these techniques measure membrane integrity alone. It has been postulated [86] that shear-related membrane damage in hybridoma cells may only be temporary and, thus, is not necessarily indicative of viability loss. Recent bursting studies with algal *Nitella syncarpa* cells [87] have revealed the ability of the cell to seal wounds in the plasmalemma. Furthermore, due to adverse effects on other metabolic activities, viability may be lost at shear stress levels below those necessary to cause membrane damage. In a number of recent plant suspension studies [56–58] membrane integrity has been shown to be a less sensitive indicator than other physiological and metabolic quantities.

4.3
Release of Intracellular Components

Cell response to hydrodynamic stress has frequently been evaluated in terms of the release of intracellular components into the suspending fluid. An increase in the extracellular concentration is not necessarily indicative of cell lysis; in filamentous fungal and bacterial systems it was postulated that the leakage of low molecular weight nucleotides might reflect variations in cell strength [88]. In shear studies of plant suspensions, extracellular phenolics [77, 89] and protein [58] have been monitored. Variations in suspension pH have been associated with the release of a secondary metabolite [90]. Release of lactate dehydrogenase (LDH), commonly used as an indicator of lysis in animal cells (e.g. [86, 91]), is less applicable to plant cell systems because of the very low prevailing LDH concentrations [92].

4.4
Morphological Variations

Variations in the size, shape and aggregation patterns of plant cells are commonly associated with the hydrodynamic environment, although the morphological effects are seldom as dramatic as those observed in filamentous fungal systems. Broth rheological properties may also be affected by morphological characteristics, with increased cell sizes resulting in higher apparent viscosities [23–25, 93]. For plant cell suspensions, some degree of cell–cell contact is regarded as essential for secondary metabolism [94]. A number of studies have pointed to a strong relationship between aggregate size and metabolite production (e.g. [45, 95–98]). Such effects have been variously associated with some degree of cellular organization or differentiation [94], a reduction in light intensity along the radius of the cell aggregate [97], or diffusional limitation in larger aggregates [96]. For oxygen limitation, a critical diameter of about 3 mm has been suggested [96, 99]. More recently, however, Keßler et al. [99] reported that in suspensions of *C. roseus*, specific ajmalicine production (based on active biomass, excluding stored carbohydrates) was independent of aggregate size for all aggregates of less than 250 μm in diameter. Higher metabolite production levels in these aggregates were identified as a beneficial effect of eddy-based hydrodynamic stress. An increase in the intensity of the cultivation environment, specifically transferred from shake flask to bioreactor, has been variously reported to increase [100] and reduce [55] average aggregate size.

For the analysis of the effects of hydrodynamic stress, plant cell aggregates can be separated into different size fractions by sieving. Image analysis [53, 55, 101] facilitates more comprehensive sample analysis and is of particular benefit in the study of particles with a narrow size distribution range, less suited to sieve separation. Rosenberg [69] used scanning electron microscopy to detect shear-induced changes in cellular morphology.

The expansion index [51, 102, 103], defined as the ratio of cell fresh weight to dry weight evaluated at the time of maximum fresh weight, is a more qualitative indicator of changes in cell/aggregate size, under various cultivation conditions. Wongsamuth and Doran [58] identified the filtration characteristics of suspensions of *Atropa belladonna*, specifically cake permeability, which is at least partially related to morphology, as a useful indicator of shear effects.

4.5
Metabolic Indicators

In both plant (e.g. [57]) and animal (e.g. [86]) cell systems, cellular respiration has been shown to be a more sensitive indicator of system response to hydrodynamic stress than membrane integrity, suggesting that intracellular enzymes and/or organelles may be affected at stress levels lower than those required to cause membrane damage.

A method commonly employed for the indirect evaluation of cell viability is based on the reduction of tetrazolium salts in the mitochondria of living cells. Towill and Mazur [104] reported on the reduction of 2,3,5-triphenyltetrazolium chloride (TTC) as a viability assay for plant suspension and tissue cultures. Since then, the method has been widely applied in hydrodynamic shear studies (e.g. [29, 55–57, 69, 77, 102]). Comparison with other putative viability indicators (e.g. membrane integrity, regrowth potential) [55, 57] suggests that it reflects a more subtle cellular response. As described by Towill and Mazur [104], a number of experimental parameters, including tetrazolium salt concentration, buffer pH, and incubation time, must be optimized prior to application of the assay with a particular system. In a recent study using *Fragaria ananassa* and *Eucalyptus perriniana* suspensions [105], it was shown that formazan production was also influenced by the age and conditions of the suspension sample. The data reported undermine the general applicability of the TTC assay as a quantitative indicator of viability.

The effect of a particular cultivation environment on a system can be evaluated in terms of biomass (fresh/dry weight, cell number), secondary metabolite production [51, 75, 89, 102, 103, 106, 107] or substrate consumption (e.g. carbon source [57] or oxygen [53, 108]). Using the Evan's Blue method to identify non-viable cells, Ho et al. [108] used viable cell density measurements to determine variations in specific growth rate attributable to hydrodynamic stress.

Keßler et al. [109] correlated loss of viability in vigorously agitated strawberry suspensions with variations in acid phosphatase activity. Studies of a wide range of intracellular factors, including DNA, ATP, NAD(P)H and Ca^{2+} levels [55, 56], have helped to indicate the extent of cellular response to hydrodynamic stress. The general state of a culture can be comprehensively assessed using flow cytometry, which facilitates analysis on a cell-by-cell basis. Plant cell applications have been limited due to the morphological heterogeneity of cells in suspension and the predominance of aggregates. Yanpaisan et al. [110] reported a flow cytometric study of the suspension cell cycle properties of *Solanum aviculare* cultivated in shake flasks at different agitation intensities. All measurements were performed using suspensions of nuclei, mechanically isolated from suspended cells. Increasing the shaker speed was observed to affect the relative distribution of cells within the cell cycle. The proportion of cells in the S phase was increased, while the proportion in the combined G_0 and G_1 phases was reduced. While broadly similar effects have been reported for mammalian cells [111], no single mechanism could be employed to describe the results. Although technically very demanding, this method has enormous potential for characterizing cell populations.

5
Shear Sensitivity Studies Under Defined Flow Conditions

Studies of the shear sensitivity of biological systems can be broadly divided into two categories, classified in terms of the prevailing hydrodynamic environment. In the first category, cell suspensions are exposed to well-defined laminar or turbulent flow conditions, in purpose-built shear devices. Studies of this type involving plant cell cultures are summarized in Table 2. As these devices (Couette, capillary and submerged jet) are frequently not designed for extended sterile operation, and are thus inappropriate for cultivation purposes, the duration of exposure to shear is generally relatively short. In such cases, the properties of the cells can be assumed to be constant. With appropriately designed systems, changes in the condition of the suspension can be directly related to the intensity of the flow field and the relevant exposure time.

5.1
Capillary Flow

The capillary tube forms the basis for one of the simplest devices employed in the evaluation of shear effects on biological suspensions and has been used in studies involving human blood [112], mammalian cells [113–115], insect cells [116], enzyme solutions [117–119] and plant cells [29, 54, 120]. For a tube of radius r, measurements of pressure drop (Δp) per unit length of capillary tube (l) allow determination of the levels of shear to which the suspended particles are subjected. The range of operating conditions can be extended by modifying the viscosity of the suspending fluid with additives such as dextran or carboxymethylcellulose. With fluids of known rheological characteristics, under conditions of fully developed laminar flow ($Re < 2100$), the wall shear stress (τ_w) and the associated shear rate (γ_w) may be calculated at a volumetric flow rate (Q) using the following equations:

$$\tau_w = \frac{r\,\Delta p}{2\,l} \tag{6}$$

For a Newtonian fluid:

$$\gamma_w = \frac{4Q}{\pi\, r^3} \tag{7}$$

Assuming a parabolic velocity profile for laminar conditions, the flow averaged shear stress is calculated as:

$$\bar{\tau} = 0.533\,\tau_w \tag{8}$$

For turbulent flow, the average shear stress may be evaluated, assuming a one-seventh power law profile, as:

$$\bar{\tau} = 0.632\,\tau_w \tag{9}$$

Table 2. Defined flow field shear studies involving plant cell suspension cultures

Cell system	Apparatus	Flow regime	Response indicator	Growth/Non-Growth	References
Datura stramonium	rheogniometer	oscillatory	membrane integrity	non-growth	[31]
Morinda citrifolia	recirculating flow capillary	laminar/turbulent	membrane integrity; morphology	non-growth	[54, 120]
Daucus carota	Couette viscometer	laminar	regrowth; mitochondrial activity; membrane integrity; cell lysis (cell number)	non-growth (sterile)	[57, 69]
Daucus carota; Petunia mitchelle; Glycine soja	Haake rotoviscometer (paddle system)	turbulent	mitochondrial activity; cell lysis (cell no.); aggregate size	non-growth	[57, 69]
Morinda citrifolia	submerged jet	turbulent	membrane integrity; morphology	non-growth	[59, 129, 130]
Nicotiana tabacum	Couette-type	transitional/turbulent	cell lysis (dry weight reduction); mitochondrial activity; metabolite secretion	non-growth	[77]
Perilla frutescens	Couette-type		mitochondrial activity	non-growth	[102]
Taxus cuspidata	rotating wall vessel	laminar	biomass and metabolite productivity	growth	[106]
Pirus malus	Couette-type	laminar	biomass production (cell no.)	growth	[141]

Capillary shearing devices are generally of two main types, namely single-pass or recirculating flow. In single-pass devices, a sample flows only once through the test section and exposure times are comparatively short. Recirculating flow devices, on the other hand, allow for repeated exposure of the suspension to a given test section and, accordingly, for longer exposure times. Fluid recirculation requires a pumping device and peristaltic pumps are commonly used. Thus, control experiments are essential to establish the effect which the pumping mechanism and flow through ancillary fittings may have on the suspension. Control runs performed by simply removing the capillary test section from the loop do not accurately represent actual pumping effects as, in the absence of the test section, there is a significant reduction in the back pressure to which the system is subjected. Kieran et al. [54] have described a procedure for performing meaningful control experiments at representative operating pressures. For suspensions of *M. citrifolia*, it was determined that correction for pump damage was essential at pumping pressures in excess of 1.5 bar [120, 121].

If well-developed laminar flow is required in either single-pass or recirculating flow devices, care must be taken to quantify and minimize entrance and exit effects. This may be accomplished by the use of converging/diverging sections at the capillary inlet and outlet. Because of the risk of capillary plugging and/or size-related effects, these devices are unsuitable for use with larger particles or aggregated systems. Passage of a cell through a recirculating flow loop combines short, highly intense flow through the test section with an extended period in the comparatively benign reservoir environment. It has been postulated [116] that this qualitatively simulates the cell's experience in a bioreactor, comprising short periods in the vicinity of the impeller where local energy dissipation rates are highest, followed by circulation in the bulk of the bioreactor where mixing is less intense. Quantitatively, the energy dissipation rates prevailing in capillary shear devices (up to 25 $kW kg^{-1}$ [29]) are many orders of magnitude higher than those that might be anticipated under normal operating conditions in a bioreactor.

Capillary shear tests were performed on low density (50 g fresh weight l^{-1}) suspensions of *M. citrifolia* using the apparatus illustrated in Fig. 1. Under both laminar and turbulent conditions [54, 120, 121], the relative viability of the suspension, evaluated using the Evan's Blue dye exclusion technique, was found to fall with exposure time in the loop. Loss of viability is well described by a first-order model:

$$\frac{X}{X_0} = e^{-k\theta} \tag{10}$$

where X_0 is the pre-shear viability, X is the viability after an exposure time, θ, and k is a first-order rate constant.

Similar kinetics have been observed for some [91] but not all [116] animal/insect lines. Trials conducted over a range of average shear stresses (Fig. 2) clearly indicate a higher degree of suspension sensitivity to turbulent, rather than laminar, flow conditions. Similar effects have been reported by other workers for plant [57] and mammalian [86, 114, 122] systems. From these

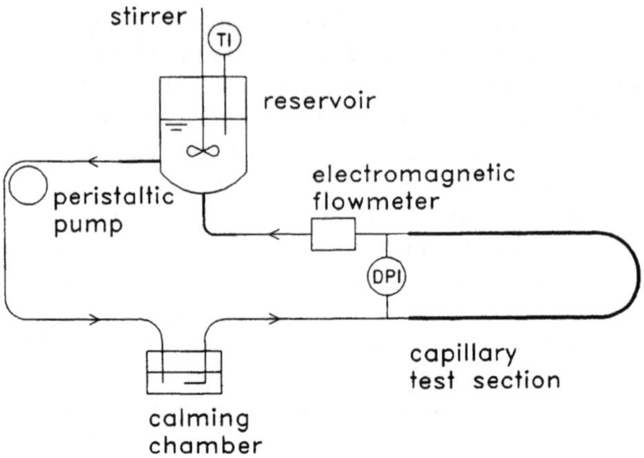

Fig. 1. Schematic diagram of the capillary flow loop apparatus. From reference [54], reproduced with permission. © 1995, Wiley-Liss, Inc, a subsidiary of Wiley, New York

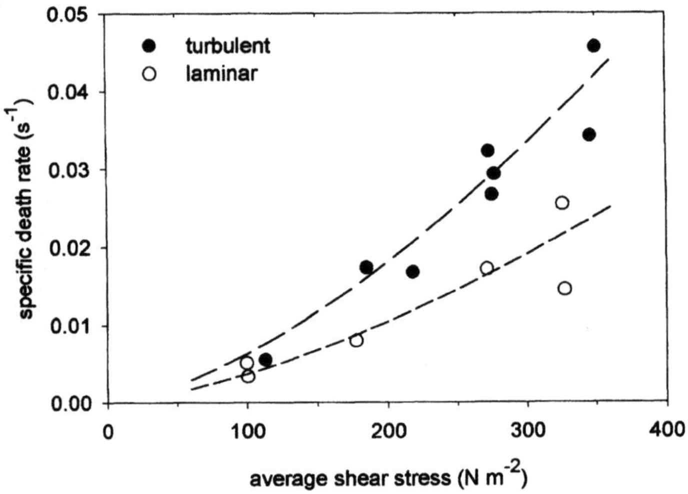

Fig. 2. Variation of specific death rate with average shear stress, under laminar and turbulent conditions, in the capillary flow loop apparatus [54, 121]

trends, it may be postulated that the mechanism for cell damage may depend on the flow regime, e.g. "stretching" under laminar flow conditions, as opposed to eddy bombardment in turbulent flow. An alternative, however, is that, under turbulent conditions, average shear stresses and energy dissipation rates are less appropriate as correlating factors for damage than locally evaluated quantities, which may be significantly higher. This approach was adopted by Kresta and Zhou [40, 43, 123] who found that mean drop size in dispersed liquid-liquid systems was better correlated with maximum energy dissipation rates than with average rates.

In capillary shear studies involving *M. citrifolia*, the extent of cell damage was found to increase with the prevailing level of shear stress (Fig. 2). Trials involving capillary tubes of different lengths yielded similar levels of viability loss at equivalent exposure times, indicating that the death rate is determined by the shear stress alone.

5.2 Jet Flow

The turbulent jet is a well-characterized flow geometry which has been used in investigations of the shear susceptibility of green algae and cyanobacteria [124], and insect cells [125]. In non-biological applications, it has been used to study the effects of mixing on the progress of very fast chemical reactions [126, 127]. The submerged jet device shown in Fig. 3 has been employed under non-growth conditions to expose suspensions of *M. citrifolia* to shear forces [59, 78, 128–130].

Flow in a submerged jet is characterized by a Reynolds number:

$$Re = \frac{\rho u_o d}{\mu} \tag{11}$$

where u_o is the nozzle exit velocity, and d is the nozzle diameter.

For values of the Reynolds number in excess of 3000, flow is assumed to be turbulent. As the fluid emerges from the nozzle, the jet expands in the flow direction (Fig. 4) by progressively entraining the surrounding quiescent fluid. For some 6 nozzle diameters downstream of the nozzle, a cone-shaped core of fluid retains the nozzle exit velocity. Within this so-called flow development region, the flow is potential and damage effects are likely to be minimal. In the expanding shear layer which forms outside of this potential core, significant turbulent stresses are generated. Downstream of the potential core, in the fully developed

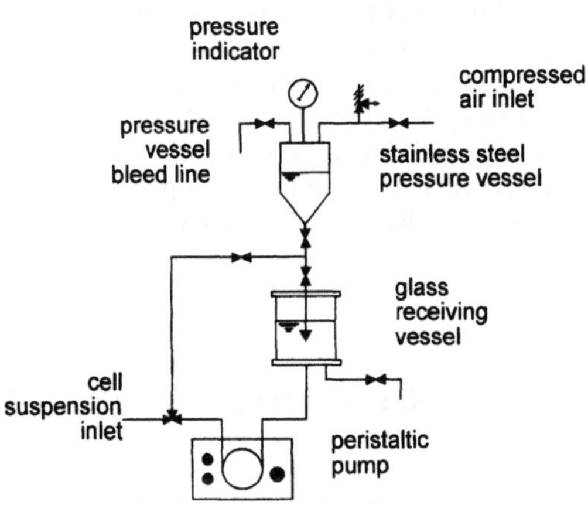

Fig. 3. Schematic diagram of the submerged jet apparatus. From reference [59], reproduced with permission. © 1998, Wiley-Liss, Inc, a subsidiary of John Wiley & Sons, Inc

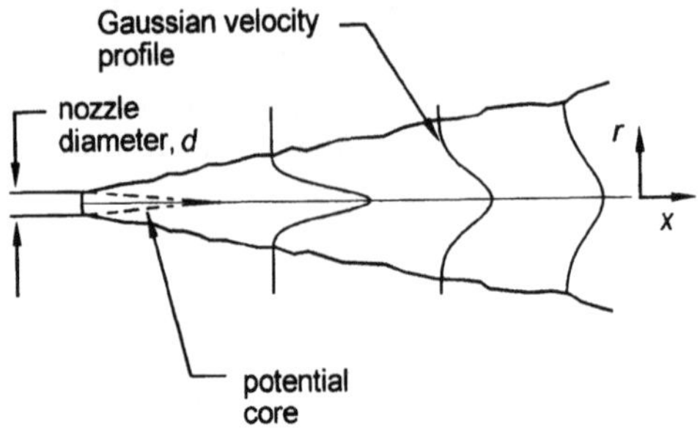

Fig. 4. Flow in a submerged jet. From reference [59], reproduced with permission. © 1998, Wiley-Liss, Inc, a subsidiary of John Wiley & Sons, Inc

region, the flow becomes fully turbulent across the expanding jet cross-section and a Gaussian velocity profile is established:

$$\frac{u}{u_m} = e^{-Ky^2} \tag{12}$$

where y is the dimensionless radial distance (r/x). The center line velocity, u_m, decreases with axial distance from the nozzle (x), according to the equation:

$$\frac{u_m}{u_o} = C\frac{d}{x} \tag{13}$$

At large distances from the nozzle, the axial velocity exhibits self-preserving similarity in that, when plotted in the above dimensionless form, velocity profiles at all cross-sections downstream collapse onto a single curve. It is probable that strict similarity is preserved only at axial distances in excess of 30–40 diameters. However, the above formulation is frequently used to describe the velocity profile at all points downstream of the flow development region. For work performed with *M. citrifolia*, the numerical values recommended by Panchapakesan and Lumley [131] for the constants K and C in Eqs. (12) and (13), i.e. 75.2 and 6.06, respectively, are employed.

In terms of shear stress levels in the developed flow field, the maximum stress [131] occurs at an axial distance between 6 and 7 nozzle diameters downstream and is given by:

$$\tau = 0.018\,\rho u_o^2 \tag{14}$$

On the basis of the flow profiles described above [Eqs. (12) and (13)] it is possible to estimate a flow-averaged shear stress for a given set of flow conditions.

It is generally agreed that energy dissipation rates in the developed flow are proportional to the quantity u_o^3/d, and that they are high close to the axis and decrease sharply with radius, falling to negligible values at dimensionless radial distances in excess of 0.2 [131, 132]. The dissipation rate along the jet axis (ε_m) has its maximum value at an axial distance of approximately 7 nozzle diameters, and then decreases markedly as indicated by the expression:

$$\varepsilon_m = \frac{C_1 u_o^3 d^3}{x^4} \qquad (15)$$

where the constant C_1 has a value of the order of 50.

The implications are that the most intensive dissipation rates are identified with a relatively confined region close to the jet axis and stretching a short distance downstream from the location of maximum dissipation. At a given axial position, the radial extent of the region is determined by integrating the velocity profile, Eq. (12), until the flow equals that emerging from the nozzle. Thus, for any given set of flow conditions, an exposure time in this region and an average dissipation rate can be determined for each cycle through the jet system.

All work involving *M. citrifolia* was conducted under non-sterile conditions using cells harvested from the late exponential phase of batch growth, when the shear sensitivity of the cells was found to exhibit least variation; during this period, average cell and chain length were essentially constant [101]. Suspensions were subjected to repeated stress cycles (i.e. passages through the submerged jet) at operating pressures of between 1 and 10 bar, with jet nozzle diameters of between 1 and 2 mm. Control tests indicated that no damage was attributable to either pumping from the receiving vessel, or pressurizing the pressure vessel. As with the capillary studies (Sect. 5.1), a first-order death model described the loss in viability as a function of exposure time. Data for different nozzle diameters and a wide range of flow conditions indicated that the damage mechanism is independent of conditions at the nozzle exit and, moreover, supports the choice of the energy dissipation rate as a meaningful measure of the destructive potential of turbulent flow [59].

Image analysis studies of pre- and post-shear chain-length distributions revealed that all distributions were well described by a log-normal relationship and that longer chains were more susceptible to damage than shorter chains. After exposure to turbulent jet flows, the mean and mode of the chain-length distributions are shifted to lower values; the occupancy of the larger size intervals is significantly reduced and the spread of the distributions is accordingly decreased [128]. On this basis, length was identified as a critical factor in determining chain susceptibility to damage [128, 129] and the maximum chain length was adopted as an appropriate distribution parameter to follow the breakage process.

A similar approach has been suggested in other studies of plant cells [57] and protein precipitates [133]. However, information on the rate of the size distribution shift process cannot be inferred from chain-length measurements made only at the beginning and end of the experiment. To date, there have been no reports on the progressive modification of the size distribution of plant cells subjected to continued exposure to turbulent forces. There are, however, a number of studies which address the break-up of mycelial hyphae in agitated vessels

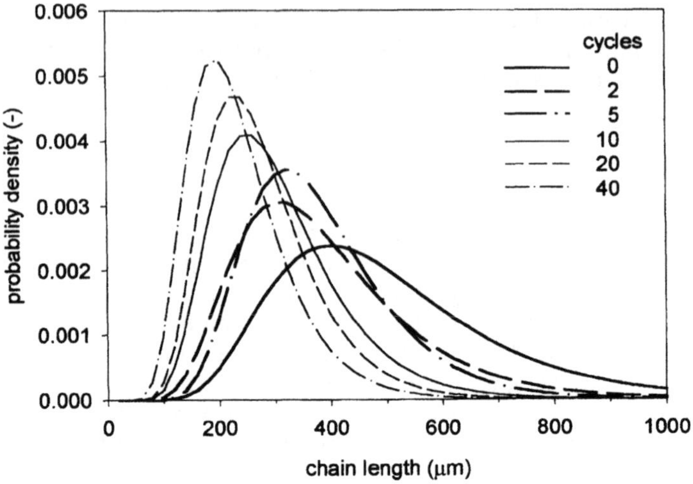

Fig. 5. Variation in the chain-length distribution in *M. citrifolia* suspensions during a typical experimental jet run [130]

(e.g. [134–136]). More recently [129, 130], the progressive chain-breakage process in suspensions of *M. citrifolia* has been followed by monitoring the chain-size distribution at several stages during the stress test. Pressures ranging from 1 to 10 bar and nozzle diameters of 1.5 and 2 mm generated jet discharge velocities in the range 15 to 39 $m s^{-1}$, with nominal energy dissipation rates of between 12 and 170 $kW kg^{-1}$.

Population balance techniques were used to model the break-up of chains of *M. citrifolia* under these conditions. Figure 5 shows the typical variation in the (log-normal) fitted chain-length distribution during the course of a representative experimental run. Experimental data points are omitted for clarity. This trend suggests that the largest chains are preferentially disrupted and that breakage proceeds until a relatively more resistant population, consisting predominantly of chains with two or three cells, remains. Chains which were observed on microscopic examination of samples stained with Evan's Blue to contain one or more non-viable cells were defined as "non-viable" chains. Chain death rates were calculated by applying a first-order death model to chain viability measurements, and the data presented in Fig. 6 indicate a strong positive correlation with average energy dissipation rate [130].

Applying population balance methods to the evolving chain-length distribution suggested that breakage rates are linearly dependent on chain length. Furthermore, investigation of a range of breakage kernels suggested that the most probable type of chain breakage is that which produces two "daughter" chains of approximately equal lengths [130].

In comparing the capillary- and jet-based studies of *M. citrifolia*, it is important to note the differences between the two experimental systems. As employed, the capillary geometry generates flow fields with moderate dissipation rates (~25 $kW kg^{-1}$) and relatively long exposure times. In jet flows, cells are ex-

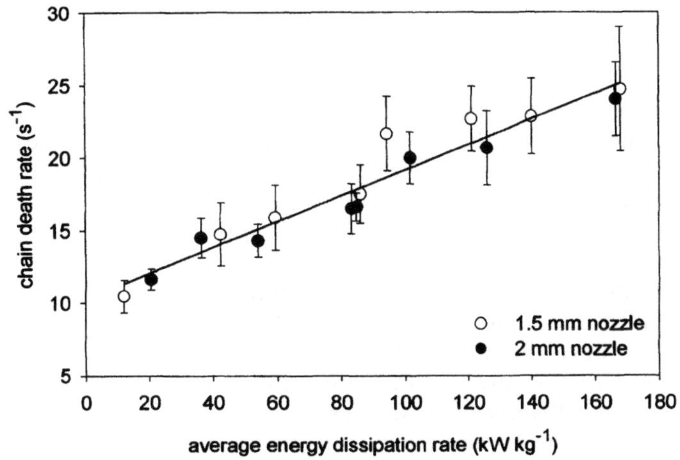

Fig. 6. Variation in chain death rates with energy dissipation rate in submerged jet flow [130]

posed to very intense turbulence ($\varepsilon < 1000$ kW kg^{-1}) for very short exposure times. However, the significance of dissipation rate as a meaningful measure of the destructive potential of turbulent flow, independent of system geometry, is clearly indicated in Fig. 7, where first-order cell death rate constants for capillary and jet flow are presented. The offset between the fitted lines is entirely attributable to the choice of active volume for energy dissipation.

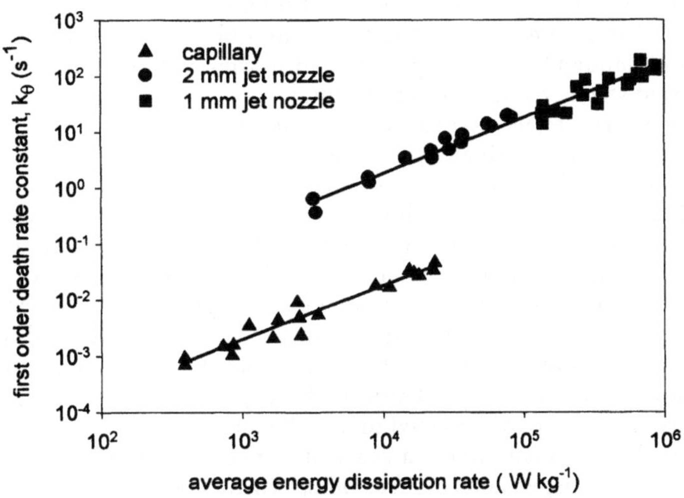

Fig. 7. Variation in cell death rates with average energy dissipation rate for capillary and jet flows. From reference [59], redrawn with permission. © 1998, Wiley-Liss, Inc, a subsidiary of John Wiley & Sons, Inc

5.3 Couette Flow

In shear studies, the most commonly used type of device for the generation of well-defined flow fields is the rotational viscometer. The use of these devices for the rheological characterization of liquids is well established [137]. Compared with the capillary and jet devices (Sects. 5.1 and 5.2), rotational viscometers allow the investigation of the effects of continuous rather than intermittent shearing.

Cone-and-plate viscometers have been employed to study shear effects in both suspended (e.g. [138]) and anchorage dependent [122] mammalian cells. These devices have the advantage of requiring only small sample volumes (~1 ml). However, they are generally inappropriate for plant cell suspensions due to the larger cell and aggregate sizes.

In co-axial or Couette viscometers, the gap size, usually less than 10% of the cylinder diameter, may restrict applications to smaller suspended particles such as freely suspended mammalian or unicellular bacterial systems. Aggregates may be pre-screened to exclude larger particles [57, 69], although this approach may preclude the use of a representative sample. Other issues which must be considered during extended operation are cell settling and evaporation from the open liquid surface of the viscometer. The latter may be particularly relevant if surface effects are known to be significant, as is the case for enzymes [139]. End effects, the existence of non-uniform shear fields in the regions above and/or below the rotating element, commonly occur with simpler cup-and-bob devices. To account for these effects, Hooker et al. [77] assumed a model for flow in the non-annular region and estimated its relative contribution to the total shearing effect. Alternatively, the rotating bob may be modified to generate well-defined flow conditions throughout the surrounding volume (e.g. [140]). The commercial rheometers commonly employed in shear studies are seldom intended for sterile operation. In studies with plant cells, Rosenberg [69] achieved sterility with alcohol washings.

For short-term trials (<12 h) conducted under non-growth conditions, Hooker et al. [77] used a single-gap, Couette-type device to expose suspensions of *N. tabacum* to a high shear environment at nominal shear rates of between 600 and 1200 s^{-1}. Laminar conditions did not prevail. Based on measurements of mitochondrial activity, cell lysis and secondary metabolite production and secretion, damage was found to increase with both shear rate and exposure time; suspensions in the later exponential stage of growth were observed to be particularly sensitive. More recently, Zhong et al. [102] have used a broadly similar rotating drum device, operated at nominal shear rates of up to 1000 s^{-1}, for exposure times of up to 20 min, with suspensions of *Perilla frutescens*. Mitochondrial activity was affected under all conditions.

Soule et al. [141] constructed a sparged, concentric cylinder bioreactor for the cultivation of suspensions of *Pirus malus*. Growth was reduced under all rotational conditions. Sun and Linden [106] employed a rotating wall vessel (Rotary Cell Culture System, Synthecon, Houston, TX, USA) to cultivate suspensions of *Taxus cuspidata* under laminar flow conditions. Shear rates were varied by independently adjusting the rotational speeds of the inner and outer cylinders, which were orientated horizontally. System oxygen demands were satisfied by bubble-free oxygenation through a gas-permeable membrane. Shear rates of up to $10.4\ s^{-1}$ corresponded to a maximum shear stress of $11.1\ mN\,m^{-2}$. These conditions are significantly more benign than those typically employed for testing shear sensitivity, yielding maximum cumulative energy dissipation levels of $1\times10^4\ J\,m^{-3}$ over a 14 d period. The batch trials conducted in this system indicated that Taxol accumulation was reduced at higher shear conditions ($\tau > 2\ mN\,m^{-2}$), although specific growth rates were enhanced.

Dunlop and co-workers [57, 69, 142] performed comprehensive studies of the effects of both laminar and turbulent conditions (for exposure times of up to 1 and 4 h, respectively) in rotational viscometers on plant cell suspension cultures. First-order decay kinetics [69] were found to adequately describe the reduction in membrane permeability with time. Some deviations, particularly during the early stages of exposure, suggested that the population might be heterogeneous with respect to shear susceptibility. Other workers have reported similar effects [54, 108]. A wide range of stress indicators was employed and this work clearly identified a hierarchy of cellular responses, as well as emphasizing the difficulties associated with comparing results from different studies, without due regard to the analytical techniques employed. From Fig. 8 [57], where relative biological activity under laminar flow conditions is presented as a function of cumulative energy dissipation, it is apparent that the ability of a cell to divide is affected at total energy dissipation levels many orders of magnitude below those required to cause detectable levels of damage to the cell membrane

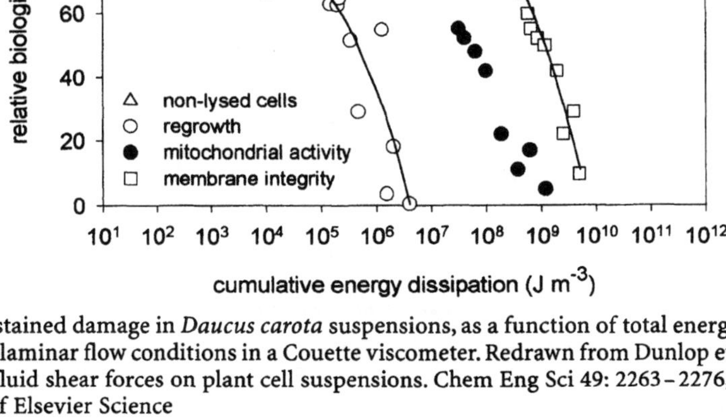

Fig. 8. Sustained damage in *Daucus carota* suspensions, as a function of total energy expended, under laminar flow conditions in a Couette viscometer. Redrawn from Dunlop et al. (1994) Effect of fluid shear forces on plant cell suspensions. Chem Eng Sci 49: 2263–2276, with permission of Elsevier Science

or cell wall. Similar trends have been reported in shear studies in STR systems [55, 58].

5.4
Comparison

Figure 9 compares, on the basis of cumulative energy dissipation, the results of turbulent flow in capillary [54], jet [59] and Couette flow [57] devices for two different cell suspensions (*M. citrifolia* and *D. carota*). There is a good level of agreement between the response trends of both biological systems to similar levels of cumulative energy dissipation, independent of morphological differences between them and of the geometry, mode and duration of operation of the shearing devices. While the viability results for the capillary trials suggest that the cells are more robust, the apparent differences are attributable to the choice of the active volume and the calculation of the average energy dissipation rate [59] which overestimates the severity of the environment in the capillary.

Even for well-defined flow fields, it is difficult to compare results collected under different conditions, for a variety of physiologically and morphologically distinct cell lines, using any of a number of stress indicators. Although valuable for a given study, first-order decay rates are inappropriate in situations where the monitored biological response may be alternately stimulated and suppressed by shear exposure. For example, from Fig. 8 it is apparent that laminar flow conditions had both stimulating and damaging effects, evaluated in terms of mitochondrial activity. The concept of a critical shear stress, causing a specified

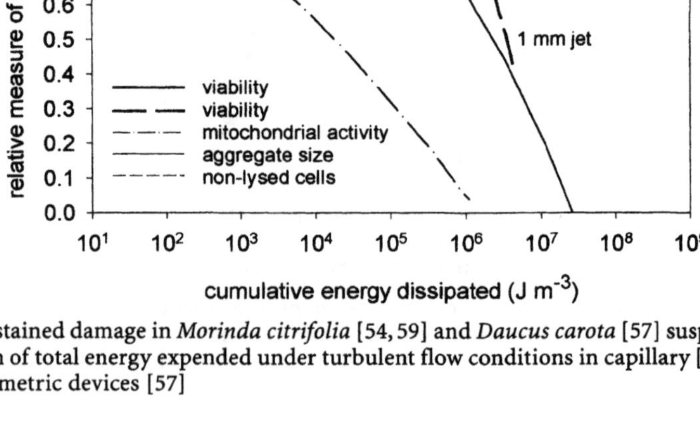

Fig. 9. Sustained damage in *Morinda citrifolia* [54, 59] and *Daucus carota* [57] suspensions, as a function of total energy expended under turbulent flow conditions in capillary [54], jet [59] and viscometric devices [57]

level of damage over a given exposure time, has also been employed by a number of workers (e.g. [54, 69, 90]). Analogous to the critical shear stress, Dunlop et al. [57] identified a critical value of cumulative energy dissipation (E_c), which resulted in a specified loss of biological activity. From Fig. 9, for mitochrondrial activity in *D. carota* suspensions, the E_c level at which the system retains 50% of its original activity is approximately 10^4 J m^{-3}; for membrane integrity of *M. citrifolia* suspensions in the jet, the corresponding value is about 5×10^6 Jm^{-3}.

6
Shear Sensitivity Studies in Stirred Tank Reactor Configurations

In these studies, suspensions are subjected to shear forces in cultivation vessels designed for extended sterile operation. Exposure may be for the duration of cultivation, or a significant portion thereof. Due to the variety of bioreactor configurations employed, the results tend to be highly system-specific. However, it can be argued that analysis of a culture under actual growth conditions, in an appropriately scaled down system, provides more potentially valuable data on scale-up prospects. Chemostat cultures facilitate the investigation of long-term effects without reference to phase-related variations [57, 78, 107, 143], but are generally of less relevance to actual production conditions, particularly where the metabolite of interest is non-growth associated. In bioreactor-based shear studies, regulation of the hydrodynamic environment is typically effected via the rate or method of agitation and/or aeration. In shake flasks, shaker speeds may be varied or baffled flasks employed. The intensity of the flow field may be characterized in any of the ways outlined in Sect. 3. A summary of recent shear-related studies in these comparatively poorly defined hydrodynamic environments is presented in Table 3. Studies prior to 1993 have been tabulated by Meijer et al. [17].

Early studies focused on the feasibility of scale-up in different reactor configurations and on the gross effects of the hydrodynamic environment. In work with *M. citrifolia*, *Beta vulgaris* and *C. roseus*, Wagner and Vogelmann [100] found that even mild mechanical agitation was detrimental to culture performance. However, it was suggested that the low agitation rate employed provided sub-optimal mixing, particularly at high biomass concentrations. Shear susceptibility was shown to be line dependent. It was also suggested that suspensions exhibited heightened shear sensitivity in the later stage of a normal batch growth cycle although, under the conditions investigated, it is difficult to conclusively isolate shear-related effects. Age/phase-related variations in shear susceptibility have been observed by a number of workers [53, 77] and are commonly attributed to changes in cell/aggregate size [54]. In modeling the kinetics of cell cultures, the tendency for cells to rupture has been associated with increased vacuolar mass [144], which also increases with cell age. Hong et al. [145] compared the performance of *F. ananassa* cv Brighton in shake flasks, an ALR, an STR and roller bottles. Operating at the minimum agitation rate required for particle suspension, the STR did not sustain growth. Poor mixing and associated oxygen deprivation may have accounted for the failure of these suspensions to grow and for the presence of significant cellular debris in the STR. More re-

Table 3. Shear studies (post-1993) involving plant cell suspensions in STR or shake flask systems. Studies prior to 1993 are summarized by Meijer et al. [1993]

Cell system	Shear system	Conditions	Response indicator	References
Stizolobium hassjoo	125-ml shake flasks	140–220 rpm; (0.4–1.7 $m^2 s^{-3}$)	secondary metabolite production; aggregate size	[45]
Cynara cardunculus; Centaurea calcitrapa	2-l STR (1.5 l wv)	N: 0.5–1.667 s^{-1}; Q_g: 0.1–0.4 vvm; t: 10–20 d; k_La: 6–24 h^{-1}	biomass production; membrane integrity; secondary metabolite production; O_2 uptake rates	[52]
Fragaria ananassa	1-l STR (1 l wv); D: 0.045 m (turbine) D: 0.05 m (anchor)	P/V: 19–1450 Wm^{-3}; Q_g: no aeration during shearing; t: < 5 h	specific O_2 uptake rate; aggregate size distribution	[53]
Morinda citrifolia	10-l STR (6 l wv); D: 0.07 m	batch and repeated batch; N: 3.33–8.33 s^{-1}; Q_g: 0.5 vvm; t: < 28 d	biomass production; membrane integrity; morphology; sensitivity to capillary flow	[54]
Carthamus tinctorius; Eucalyptus perriniana	0.55-l STR; D: 0.05 m	N: 3.33–8.33 s^{-1}; surface aeration; t: 24 h	regrowth; membrane integrity; mitochondrial activity; aggregate size; membrane fluidity; ATP, DNA, NAD(P)H, Ca^{2+}, F-actin	[55, 56]
Daucus carota	2-l STR (1.45 l wv); D: 0.09 m	chemostat; N: 1.67–5 s^{-1}; Q_g: 0.6 vvm; m t: 1000 h	biomass production; mitochondrial activity; membrane integrity; substrate consumption; extracellular DNA	[57]
Atropa belladonna	2-l STR (1.2 l wv); D: 0.049 m	N: 6.67–16.7 s^{-1}; Q_g: 0.08 vvm; t: 13 d	biomass production; membrane integrity; extracellular protein; aggregate size; cake filtration properties	[58]
Perilla frutescens	5-l STR (3 l wv); D_1: 0.065 m; D_2: 0.085 m	N_1: 1.17–4.17 s^{-1}; N_2: 1.5–4.17 s^{-1}; Q_g: 0.033 vvm; t: ~12 d	biomass production; substrate consumption; expansion index; secondary metabolite production	[102]

Table 3 (continued)

Cell system	Shear system	Conditions	Response indicator	References
Cathanthus roseus; Nicotiana tabacum; Cinchona robusta; Tabernaemontana divaricata	3-l STR (2 l wv); D: 4.5 cm	batch and chemostat; N: 2.5–16.7 s^{-1}; Q_g: 0.33 vvm; t: < 30 d	biomass production; total organic carbon; extracellular protein; membrane integrity; substrate consumption	[107]
Nicotiana tabacum	3-l STR (2 l wv); D: 0.045 m	N: 1.67–5.83 s^{-1}; Q_g: 0.43 vvm; t: 24 d	total biomass production; viable biomass production; specific O_2 consumption rates	[108]
Fragaria ananassa	0.23-l STR; D: 0.03 m	N: 40 s^{-1} (P/V: 28 $kW m^{-3}$); t: 5 h	membrane integrity; acid phosphatase activity	[109]
Solanum aviculare	250-ml shake flasks	100–600 rpm (3.5 cm throw); t: 10 d	cellular DNA distribution (flow cytometry)	[110]
Catharanthus roseus	3-l STR (1.8 l wv); D: 0.045 m; 3-l bubble column	N: 4.17 s^{-1}; Q_g: 0.33 vvm	biomass production; alkaloid production; off-gas composition	[164]

cently, Keßler et al. [109] presented data to suggest that *F. ananassa* cv Shikinari suspensions were particularly robust, exhibiting less than 10% loss of viability after 5 h of exposure to a high shear environment, with a specific power input of 28 kWm^{-3}; however, no information on subsequent regrowth was provided. Scragg and co-workers [51, 75, 103, 146–149] reported on an extensive series of studies of cell suspensions cultivated in STRs and ALRs. These works, and subsequent studies by Meijer [72], suggested that shear susceptibility is not only cell line dependent, but is also related to culture age, history and maintenance conditions. Cultures may become "adapted" to growth under high shear conditions. For example, Tanaka et al. [150] suggested that the observed robustness of a particular *C. roseus* line might be attributable to its ability to respond to the hydrodynamic environment by accelerating or adjusting cell wall synthesis. Alternatively, aggregates in a size-based sub-population, which is more susceptible to damage, may be preferentially disrupted. Takeda et al. [55] reported a reduction in average aggregate size in *C. tinctorius* suspensions after transfer from a shake flask to an STR. Increased biomass levels at higher agitation rates were attributed to an associated increase in the specific surface area.

In order to investigate the response of an exponentially growing population of cells to extended exposure to an STR environment, Kieran et al. [54, 101] cultivated *M. citrifolia* suspensions in repeated-batch or "draw-fill" mode. Although viability was in excess of 90% at all stages, there was a reduction in both biomass and metabolite productivity. The draw-fill cultivated suspensions were shown to exhibit increased sensitivity to turbulent flow in a capillary loop [54]. This effect was only partially attributable to increased chain lengths. Ho et al. [108] focused on loss of viability in suspensions of *N. tabacum*. A reduction in specific growth rate, calculated on the basis of viable cell density, indicated system damage as agitation rates were increased from 100 to 325 rpm. An increase in specific oxygen consumption rates also pointed to increased maintenance energy requirements in the stressed suspension.

Recent STR-based studies have tended to focus on sub-lethal effects, as evidenced by the increase in the number and diversity of the response indicators employed (Table 3). Dunlop et al. [57] grew carrot suspensions in chemostat culture, with step-changes in agitation rate. Steady-state biomass levels were reduced at impeller speeds in excess of 200 rpm; complete washout occurred at 500 rpm. In terms of steady-state biomass levels, system performance was reduced at a speed of 300 rpm. Overall, the results were indicative of growth inhibition, rather than cell lysis, and also suggestive of increased maintenance energy costs. Cumulative energy dissipated in this chemostat study varied between 10^3 and 10^5 Jm^{-3} and, over this range, complete loss of biological activity was observed. This represents a more dramatic effect, at significantly lower energy dissipation levels, than observed in the accompanying viscometric studies (Fig. 8) and further emphasizes the difficulty of extrapolating from short-term shear studies conducted under non-growth conditions, albeit in a well-defined flow field.

Wongsamuth and Doran [58] exposed batch-cultivated suspensions of *A. belladonna* to energy dissipation levels in the range 10^6–10^9 Jm^{-3}. Here again, a hierarchy of responses to hydrodynamic stress was identified, with membrane integrity being the least sensitive indicator. Based on cell cake filtration characteristics, cake permeability, which is strongly, but not uniquely, correlated with aggregate morphology, was also identified as a useful indicator of shear effects. In shorter term (24 h) studies with eucalyptus [56] and safflower [55, 56], Takeda et al. concentrated on the elucidation of various aspects of the process by which mechanical (hydrodynamic) stimuli are transduced and communicated within plant cells. A very wide range of metabolic indicators was monitored in response to energy dissipation rates of up to 2.3 $W kg^{-1}$. These conditions correspond to cumulative energy dissipation levels of 2×10^8 $J m^{-3}$. Membrane integrity losses were less pronounced than reductions in ATP content and respiration activity. NAD(P)H levels were reduced and a key role for cytosolic Ca^{2+} in stress response was identified (Sect. 8.1). Membrane fluidity was also measured during shearing trials [55]. Although work with animal cells has suggested that shear sensitivity might be associated with membrane fluidity [151], tests performed on safflower cells (rather than protoplasts) revealed no significant change in the safflower suspensions, over the 24-h shearing period.

7
Interfacial and Aeration Effects in Plant Bioreactors

While the parameters outlined in Sect. 3 to describe the hydrodynamic environment in bioreactors focus almost exclusively on fluid–particle interactions, most bioreactors contain three-phase systems, in which the effect of gas bubbles must also be considered. Although extended oxygenation of plant systems has been accomplished through surface aeration [152] and the use of fluorocarbon-polymer films [153], it is most commonly achieved via sparging. The sensitivity of animal and insect cell cultures to damage associated with gas bubbles has been comprehensively reviewed by Chalmers [19]. Studies of bubble–cell interactions in pneumatically impelled systems have identified the bubble disengagement process as a source of significant cell damage. Cells adhering to the gas–liquid interface are killed by the hydrodynamic forces associated with bubble rupture. However, addition of protective agents, such as the non-ionic surfactant Pluronic F-68, which are thought to prevent cell adherence at the interface, has been reported to reduce cell damage. Although preliminary studies on the effects of Pluronic F-68 on shake flask suspensions of *Solanum dulcamara* [154] indicated biocompatability at low concentrations ($<1\%$), no data are available for sparged conditions.

In contrast to the numerous studies of interfacial effects on animal cells, there is almost no comparable information for plant cell systems, implicitly suggesting that the problem may not be as significant. Because, in suspension culture, plant cells generally occur in aggregates, interactions between bubbles and single cells would be expected to be less significant than in mammalian cultures. In foaming and cell flotation studies with suspensions of *A. belladonna*, Wongsamuth and Doran [155] reported viabilities of 90–100% in cells concentrated in a foam layer, which was collapsed for subsequent recovery and analysis. Singh and Curtis [156] reported successful callus formation from cells ejected from the liquid surface of a bubble column by bubble bursting.

However, there are no detailed studies of plant cells during more controlled bubble-bursting events. Garcia-Briones et al. [157] estimated that maximum energy dissipation rates of the order of 10^7 $J m^{-3} s^{-1}$ were associated with the rupture of a 1.77-mm bubble, occurring during a 0.002-s period. On this basis, the cumulative energy dissipated during a single event is of the order of $10^4 J m^{-3}$. For a 6-mm bubble, the corresponding value is of the order of $10^3 J m^{-3}$. These levels are significantly below those reported to cause cell lysis, or loss of membrane integrity, in plant cells in turbulent flow (Fig. 9). A cell which becomes entrapped in a foam layer may, of course, be exposed to a number of bubble-bursting events. However, cells entrained in the thick foam meringue, which is common in plant cell suspensions cultivated without foaming agents in aerated bioreactors, are already lost from the active broth, unless measures to frequently resuspend the meringue are implemented.

Ballica and Ryu [158] correlated reductions in cell yield in *Datura stramonium* suspensions with the increased Reynolds stresses associated with higher aeration rates in a 1.2-l ALR. A more recent study [159] of *C. roseus* suspensions cultivated in a 1.5-l bubble column showed that the increased bubble sizes associated with both larger sparger pores and higher aeration rates caused a reduction in system performance. Here, also, it was postulated that the effects were due to increased Reynolds shear stresses in the flow field. However, it was not possible to rule out gas-stripping effects.

Other non-shear effects associated with aeration have been found to be inhibitory in plant cell systems, particularly stripping of CO_2 and essential volatiles at higher gas flow rates (e.g. [160–163]). This phenomenon was clearly demonstrated by Schlatmann et al. [164], working with *C. roseus*. By recirculating exhaust gas from an STR, ajmalicine production profiles similar to those observed in shake flasks were obtained. This work emphasized the importance of dissociating shear- and aeration-related effects. Smith et al. [165] described a strategy for independently controlling dissolved O_2 and CO_2 concentrations under conditions of constant shear. This approach was subsequently adopted by Zhong et al. [102] for controlling air, CO_2 and O_2 flow rates at a constant total gas flow rate, and goes some way towards overcoming the fundamental limitation of shear studies performed under conventional operating conditions. However, it does not address other components, as yet unidentified, which might be stripped from the system.

8
A Biological Basis for Cell Response to Shear Effects

A cell subjected to a stress of any kind can potentially exhibit a wide range of responses. Severe stress may lead to cell death and, ultimately, to cell lysis; imposition of less severe conditions may result in a metabolically perturbed system, which may either revert to its initial state or adapt in some way to the imposed conditions. Figure 10 shows a hypothetical scheme, presented by Prokop and Bajpai [12], for the signal-response cascade associated with hydrodynamic shear stress. The signal reception/transduction mechanisms are, as yet, poorly understood. While Fig. 10 can be applied to any biological system, Namdev and

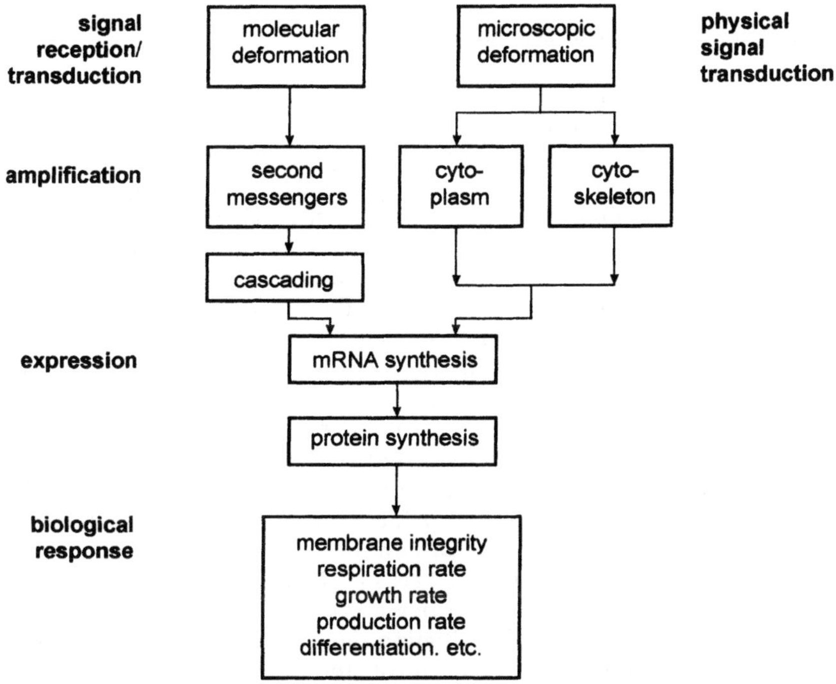

Fig. 10. Hypothetical scheme of the signal-response cascade for hydrodynamic shear stress [12]. Redrawn with permission of Academic Press, Inc.

Dunlop [18] proposed a model for sub-lytic effects in plant cells, based on the same principles, but including four properties postulated to be of particular importance in these systems, namely calcium ion flux, osmo-regulation, cell–cell contact/aggregation and stress protein expression. Of these factors, osmo-regulation (and its inter-relationship with the cell wall) and aggregation patterns, in particular, distinguish plant cells from mammalian cell systems.

Use of cumulative energy dissipation levels as a correlating parameter for cell damage is very useful, insofar as it permits the broad identification of conditions likely to elicit a particular response in a cell culture, while recognizing that there may be system-specific variations in behavior. However, it does not elucidate a mechanism for cell damage. Recent work, focusing on sub-lytic responses in plant cells, stimulated by stresses of various kinds, has gone some way towards identifying key processes in stress perception and stress responses and may provide a framework for future study in this area. One topic which has attracted considerable interest in plant and plant cell research in recent years is the phenomenon known as the oxidative burst. It illustrates elements of the model shown in Fig. 10. It incorporates the concepts of ion fluxes and osmo-regulation highlighted by Namdev and Dunlop [18] and may go some way towards explaining cellular response to hydrodynamic shear.

8.1
The Oxidative Burst in Plant Cell Suspensions

Plants have developed an array of defense mechanisms which afford them protection against pathogenic attack. These responses, which may serve either to strengthen the host or repel the invading pathogen, include synthesis of phytoalexins, reinforcement of cell walls, production of lytic enzymes and expression of antimicrobial proteins. These defense mechanisms may be accompanied by localized plant cell death, known as the hypersensitive response (HR). However, one of the first measurable responses to infection, occurring within a few minutes of stress imposition, is the production and release of active oxygen species (AOS), including the superoxide radical, $^{\bullet}O_2^-$, the hydroxyl radical, $^{\bullet}OH$, and hydrogen peroxide (H_2O_2). This phenomenon is known as the oxidative burst (OB) and is well documented in both mammalian cells and whole plants. It was first observed in potato cells in response to infection by *Phytophthera infestans* [166]. Non-stressed cells produce small quantities of AOS. However, at higher concentrations, these AOS are potentially toxic to both the pathogen and the plant cell itself. While most AOS are unstable, intracellular H_2O_2 concentrations of up to 1 M have been reported to accumulate in a matter of minutes in fungally infected cells [167]. The OB has been implicated as a factor in the HR [168] with nitric oxide as a collaborator [169] and, overall, the OB has been identified as an initiating signal for a series of metabolic defense mechanisms [170, 171]. To date, most studies of the OB have focused on stimulation by microbial pathogens or elicitors and a number of recent reviews address this topic [171–175].

From a shear sensitivity perspective, it is particularly interesting to note that the OB has also been observed to occur in response to other, non-pathogenic stimuli, including wounding [176], agitation [177–180], pressure [181, 182] and osmotic stress [180, 183]. Relevant studies are summarized in Table 4.

A number of models have been proposed for generation of the OB and have been discussed by Bolwell and Wojtaszek [184]. One of the earliest, outlined by Apostol et al. [170] and elaborated by Mehdy [185], has been developed by Low and Merida [174] to incorporate a mechanical stimulant. Relevant elements of this model are shown in Fig. 11. The mechanism by which the mechanical signal is transduced is still unclear; however, the central role of Ca^{2+} is obvious from Fig. 11. Calcium is recognized as an important intracellular messenger in plants [186] and mechanical stimuli have been observed to result in transient increases in cytosolic Ca^{2+} levels [187]. Variations in calcium levels have, furthermore, been reported to effect changes in the organization and tension of the actin filament network in soybean suspension [188]. Supporting evidence for this effect was presented by Takeda et al. [56] who observed a reduction in F-actin content in response to hydrodynamic stress, coincident with increased cytosolic Ca^{2+} concentration. The central role of Ca^{2+} has been identified in mechanically stressed plants [187], plant cells [180, 189] and mammalian/insect cells [116, 190, 191]. Stretch-activated anion channels have been reported in plant cells [192] and have been implicated in the OB [180]. If these channels are activated by mechanical forces, there is the potential for mediation of both the OB

Table 4. Studies showing evidence of an oxidative burst in plant systems, stimulated by sources other than a pathogen or elicitor

System[a]	Stimulant	Detection method	Ref.
Bean hypocotyls (*Phaseolus vulgaris*)	incision wounding of hypocotyls	oxidative cross-linking of cell wall proteins	[176]
Soybean (*Glycine max* Merr. cv Kent)	mechanical agitation in the presence of PGA elicitor	pyranine fluorescence	[178]
Eucheuma platycladum (thalli segments)[b]	breaking of thalli; suspension agitation with magnetic bar	luminol dependent chemiluminescence	[179]
Tobacco (*Nicotiana tabacum*)	osmotic stress, baffled agitation	scopoletin fluorescence	[180]
Tomato (*Lycopersicon esculentum*)	pressure (50 mPa)	pyranine fluorescence	[181]
Vitis vinifera	pressure (50 mPa, 100 mPa)	pyranine fluorescence	[182]
Soybean (*Glycine max*)	direct application of pressure to slide-mounted cells; osmotic stress	pyranine fluorescence; 2′,7′-dichlorofluorescein diacetate	[183]

[a] Unless otherwise indicated, studies were performed using cell suspension cultures.
[b] Red macroalga.

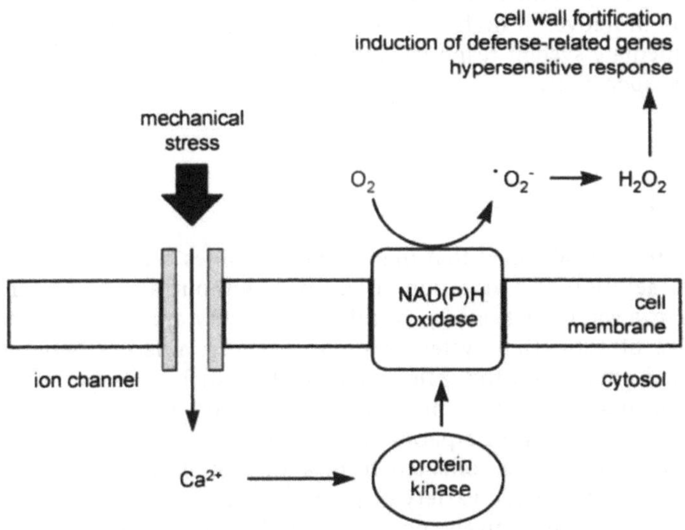

Fig. 11. Hypothetical model for the oxidative burst in plant cells, in response to mechanical stress. Based on [174] and [172]

pathway and activation of other cellular responses, as proposed by Namdev and Dunlop [18]. Work by Yahraus et al. [183], involving a stretch-activated channel blocker to inhibit the OB, was inconclusive and suggested that other channels may be involved.

The active oxygen species generated are also implicated in the oxidative cross-linking of hydroxyproline-rich glycoproteins (HRGPs) in plant cell walls in both whole plants and suspension-cultured cells. In soybean cells treated with a fungal or chemical elicitor [176], insolubilization of HRGPs was initiated within two minutes of elicitation and completed within ten. It was proposed that this process resulted in a toughening of the cell wall, as an initial stage of defense against microbial invasion. In support of this proposition, increased resistance of these cells to protoplasting enzymes was noted. This cross-linking has been widely reported in elicited cells (e.g. [176, 193, 194]) and, more recently, in response to mechanical stimulation of tobacco cells [180]. HRGPs are known to contribute significantly to the tensile strength of the cell wall [195] and it is reasonable to expect that variations in this tensile strength might affect the sensitivity of individual cells in suspension culture. Changes in cell wall composition in response to stress are not limited to HRGPs. Tanaka et al. [150] reported changes in the relative cellulose and hemi-cellulose levels between *C. roseus* suspensions cultivated in baffled and unbaffled flasks.

Work on the OB has clearly demonstrated that plant cell response to stress, including hydrodynamic stress, is a more complex process than was originally anticipated. Moreover, it highlights a number of cellular processes which may define immediate system response and longer term performance prospects of plant cell suspensions.

9
Conclusions

It has now been established that many plant cell lines can be cultivated under reasonable aeration and agitation conditions, without any significant loss of viability. However, reductions in overall productivity are common and are representative of sub-lethal system responses. Discrimination between purely shear-related effects and the influence of factors such as aeration is difficult. Although shear sensitivity has been shown to depend on both cell line and cell age, cumulative energy dissipation levels have been identified as a useful basis for comparing data collected from a variety of reactor configurations. Recent studies with a wide variety of plant cell suspensions, performed under both well- and comparatively poorly-defined flow conditions, suggest that sub-lethal responses may ultimately govern culture performance. A fuller understanding of these effects may provide the key to optimizing cultivation conditions.

References

1. Takahashi S, Fujita Y (1991) In: Komamine A, Misawa M, DiCosmo F (eds) Plant cell culture in Japan. CMC, Tokyo, p 72
2. Furuya T, Ushiyama K (1994) In: Shargool PD, Ngo TT (eds) Biotechnological applications of plant cultures. CRC, Boca Raton, p 1
3. Wu J, Zhong JJ (1999) J Biotechnol 68:89
4. Hara Y (1996) In: DiCosmo F, Misawa M (eds) Plant cell culture secondary metabolism – towards industrial application. CRC, Boca Raton, p 187
5. Sahai O (1994) In: Gabelman A (ed) Bioprocess production of flavor, fragrance and color ingredients. John Wiley, New York, p 239
6. Su WW (1995) Appl Biochem Biotechnol 50:189
7. Goldstein WE (1999) In: Fu T-J, Singh G, Curtis WR (eds) Plant cell culture for the production of food ingredients. Kluwer Academic, New York, p 195
8. Kieran PM, MacLoughlin PF, Malone DM (1997) J Biotechnol 59:39
9. Zhong JJ, Yu JT, Yoshida T (1995) World J Microb Biotechnol 11:481
10. Ritterhaus E, Ulrich J, Westphal K (1990) Int Assoc Plant Tissue Cult Newsletter 61:2
11. Curtis WR (1999) In: Fu TJ, Singh G, Curtis WR (eds) Plant cell culture for the production of food ingredients. Kluwer Academic, New York, p 225
12. Prokop A, Bajpai RK (1992) Adv Appl Microbiol 37:165
13. Hua J, Erickson LE, Yiin TY, Glasgow LA (1993) Crit Rev Biotechnol 13:305
14. Thomas CR (1990) In: Winkler MA (ed) Critical reports in applied chemistry: Chemical engineering problems in biotechnology, vol 29. Elsevier Applied Science, London, p 23
15. Joshi JB, Elias CB, Patole MS (1996) Chem Eng J 62:121
16. Merchuk JC (1991) Adv Biochem Eng Biotechnol 44:65
17. Meijer JJ, ten Hoopen HJG, Luyben KChAM, Libbenga KR (1993) Enzyme Microb Technol 15:234
18. Namdev PK, Dunlop EH (1995) Appl Biochem Biotechnol 54:109
19. Chalmers JJ (1998) Appl Mech Rev 51:113
20. Carpita N (1985) Plant Physiol 79:485
21. Taiz L (1984) Ann Rev Plant Physiol 35:585
22. Matile P (1978) Ann Rev Plant Physiol 29:193
23. Curtis WR, Emery AH (1993) Biotechnol Bioeng 42:520
24. Zhong JJ, Seki T, Kinoshita SI, Yoshida T (1992) Biotechnol Bioeng 40:1256
25. Ballica R, Ryu DDY (1993) Biotechnol Bioeng 42:1181
26. Jolicoeur M, Chavarie C, Carreau PJ, Archambault J (1992) Biotechnol Bioeng 39:511
27. Tanaka H (1981) Biotechnol Bioeng 23:1203
28. Kato A, Kawazoe S, Soh Y (1978) J Ferment Technol 56:224
29. Kieran PM (1993) PhD thesis, University College Dublin
30. Rodriguéz-Monroy M, Galindo E (1999) Enzyme Microb Technol 10:687
31. Ballica R, Ryu DDY, Powell RL, Owen D (1992) Biotechnol Prog 8:413
32. Doran P (1993) Adv Biochem Eng Biotechnol 48:115
33. Tanaka H, Nishijima F, Suwa M, Iwanoto T (1983) Biotechnol Bioeng 25:2359
34. Shibasaki N, Hirose K, Yonemoto T, Takadi T (1992) J Chem Technol Biotechnol 53:359
35. Dubuis B, Plüss R, Romette JL, Kut OM, Prenosil JE, Bourne JR (1993) In: Nienow AW (ed) Bioreactor and bioprocess fluid dynamics, BHR Group Conference Series, vol 5. MEP, London, p 89
36. Fulzele DP, Heble MR (1994) J Biotechnol 35:1
37. Matsushita T, Koga N, Ogawa K, Fujino K, Funatsu K (1994) In: Ryu DDY, Furusaki S (eds) Studies in plant science 4: Advances in plant biotechnology. Elsevier, Amsterdam, p 339
38. Baldyga J, Pohorecki R (1998) Appl Mech Rev 51:121
39. Nienow AW (1998) Appl Mech Rev 51:3
40. Kresta SM (1998) Can J Chem Eng 76:563
41. Caulet PJC, van der Lans RGJM, Luyben KChAM (1996) Chem Eng J 62:193
42. Wei T, Willmarth WW (1992) J Fluid Mech 245:619

43. Zhou G, Kresta SM (1998) Chem Eng Sci 53:2063
44. Davies JT (1972) Turbulence phenomena. Academic Press, New York
45. Huang SY, Chen KW, Chen SY (1998) J Chin Inst Chem Engrs 29:171
46. Sumino Y, Akiyama S, Fukuda H (1972) J Ferment Technol 50:203
47. Sumino Y (1992) In: Furusaki S, Enolo I, Matsumo R (eds) Biochemical engineering for 2001. Springer-Verlag, Tokyo, p 207
48. Fujita M, Iwahori K, Tatsuta S, Yamakawa K (1994) J Ferment Bioeng 78:368
49. Sinskey AJ, Fleischaker RJ, Tyo MA, Girard DJ, Wang DIC (1981) Ann NY Acad Sci 369:47
50. Kato A, Shimizu Y, Nagai S (1975) J Ferment Bioeng 53:744
51. Leckie F, Scragg AH, Cliffe KC (1991) Biotechnol Bioeng 37:364
52. Lima Costa ME, Rapaso S, Clemente A, Pais MSS, Cabral JMS (1997) In: Nienow AW (ed) Proc 4th Intl Conf Bioreactor and Bioprocess Fluid Dynamics, MEP, London, p 27
53. Keβler M, ten Hoopen HJG, Heijnen JJ, Furusaki S (1997) Biotechnol Tech 11:507
54. Kieran PM, O'Donnell HJ, Malone DM, MacLoughlin PF (1995) Biotechnol Bioeng 45:415
55. Takeda T, Seki M, Furusaki S (1994) J Chem Eng Jpn 27:466
56. Takeda T, Kitagawa T, Takeuchi Y, Seki M, Furusaki S (1998) Can J Chem Eng 76:267
57. Dunlop EH, Namdev PK, Rosenberg MZ (1994) Chem Eng Sci 49:2263
58. Wongsamuth R, Doran PM (1997) J Chem Tech Biotechnol 69:15
59. MacLoughlin PF, Malone DM, Murtagh JT, Kieran PM (1998) Biotechnol Bioeng 58:595
60. Cherry RS, Papoutsakis ET (1986) Bioproc Eng 1:29
61. Croughan MS, Hamel JF, Wang DIC (1988) Biotechnol Prog 5:137
62. Metzner AB, Otto RE (1957) AIChE J 3:3
63. Calderbank PH, Moo-Young MB (1959) Trans IChemE 37:26
64. Calderbank PH, Moo-Young MB (1961) Trans IChemE 39:337
65. Bowen RL (1986) Chem Eng 93:55
66. van't Riet K, Smith JM (1975) Chem Eng Sci 30:1093
67. Wichterle K, Kadlec M, Zak L, Mitschika P (1984) Chem Eng Commun 26:25
68. Robertson B, Ulbrecht JJ (1987) In: Ho CS, Oldshue JY (eds) Biotechnology processes: Scale-up and mixing. AIChE, New York, p 31
69. Rosenberg MZ (1987) PhD thesis, Washington University, St Louis
70. Matsuo H, Unno H (1981) ASCE J Environ Eng Div 107:527
71. Schlichting H (1973) Boundary layer theory, 7th edn. McGraw Hill, New York
72. Meijer JJ (1989) PhD thesis, Delft University of Technology
73. DiCosmo F, Misawa M (1985) Trends Biotechnol 3:318
74. Eilert U (1987) In: Constabel F, Vasil I (eds) Cell culture and somatic cell genetics of plants, vol 4. Academic Press, San Diego, p 153
75. Scragg AH, Allan EJ, Leckie F (1988) Enzyme Microb Technol 10:361
76. Horsch RB, Jones GE (1980) In Vitro 80:103
77. Hooker BS, Lee JM, An G (1989) Enzyme Microb Technol 11:484
78. Curtin FT (1991) PhD thesis, University College Dublin
79. Gaff DF, Okong'o-ogola O (1971) J Exp Bot 22:756
80. Widholm JM (1972) Stain Technol 47:189
81. Swain D, De DN (1994) Indian J Exp Bot 32:501
82. Tucker KG, Chalder S, Al-Rubeai M, Thomas CR (1994) Enzyme Microb Technol 16:29
83. Smith MAL, Reid JF (1996) In: DiCosmo F, Misawa M (eds) Plant cell culture secondary metabolism – towards industrial application. CRC, Boca Raton, chap 3, p 53
84. Markx GH, ten Hoopen HJG, Meijer JJ, Vinke KL (1991) J Biotechnol 19:145
85. Parr AJ, Smith JI, Robins RJ, Rhodes MJC (1984) Plant Cell Rep 3:161
86. Abu-Reesh I, Kargi F (1989) J Biotechnol 9:167
87. Skobeleva OV, Ktitorova IN, Marichev GA, Lyalin OO (1996) Russ J Plant Physiol 43:511
88. Tanaka H, Mizuguchi T, Ueda K (1975) J Ferment Technol 53:18
89. Hooker BS, Lee JM, An G (1990) Biotechnol Bioeng 35:296
90. Vogelmann H, Bischof A, Pape D, Wagner F (1978) In: Alfermann AW, Reinhard E (eds) Production of natural compounds by cell culture methods. GSF, Munich, p 130

91. McQueen A, Meilhoc E, Bailey JE (1987) Biotechnol Lett 2:831
92. Jervis L, Robertson ER (1987) In: Webb C, Mavituna F (eds) Plant and animal cells: process possibilities. Ellis Horwood, Chichester, p 216
93. Zhang YH, Wang HQ, Liu S, Yu JT, Zhong JJ (1997) Biotechnol Lett 19:943
94. Lindsey K, Yeoman MM (1983) J Exp Bot 34:1055
95. Kinnersly AM, Dougall DK (1980) Planta 149:200
96. Hulst AC, Meyer MMT, Breteler H, Tramper J (1989) Appl Microbiol Biotechnol 30:18
97. Madhusudhan R, Ravishankar GA (1996) Biotechnol Lett 18:1253
98. Hanagata N, Ito A, Uehara H, Asari F, Takeuchi T, Karube I (1993) J Biotechnol 30:259
99. Keßler M, ten Hoopen HJG, Furusaki S (1999) Enzyme Microb Technol 24:308
100. Wagner F, Vogelmann H (1977) In: Barz W, Reinhard E, Zenk MH (eds) Plant tissue culture and its biotechnological applications. Springer, Berlin Heidelberg New York, p 85
101. Kieran PM, Malone DM, MacLoughlin PF (1993) Trans IChemE (C) 71:40
102. Zhong JJ, Fujiyama K, Seki T, Yoshida T (1994) Biotechnol Bioeng 44:649
103. Leckie F, Scragg AH, Cliffe KC (1991) Enzyme Microb Technol 13:296
104. Towill LE, Mazur P (1975) Can J Bot 53:1097
105. Keßler M, Furusaki S (1997) J Chem Eng Jpn 30:718
106. Sun X, Linden JC (1999) J Ind Micro Biotechnol 22:44
107. Meijer JJ, ten Hoopen HJG, van Gameren YM, Luyben KChAM, Libbenga KR (1994) Enzyme Microb Technol 16:467
108. Ho CH, Henderson KA, Rorrer GL (1995) Biotechnol Prog 11:140
109. Keßler M, Bye Ø, Furusaki S (1997) Biotechnol Lett 19:545
110. Yanpaisan W, King NJC, Doran PM (1998) Biotechnol Bioeng 58:515
111. Lakhotia S, Bauer KD, Papoutsakis ET (1992) Biotechnol Bioeng 40:978
112. Beissinger RL, Laugel RF (1987) AIChE J 33:99
113. Augenstein DC, Sinskey AJ, Wang DIC (1971) Biotechnol Bioeng 13:409
114. McQueen A, Bailey JE (1989) Biotechnol Lett 8:531
115. Zhang Z, Al-Rubeai M, Thomas CR (1993) Biotechnol Bioeng 42:987
116. Aloi LE, Cherry RS (1994) J Biotechnol 33:21
117. Charm SE, Wong BL (1970) Biotechnol Bioeng 12:1103
118. Thomas CR, Dunhill P (1979) Biotechnol Bioeng 21:2279
119. Reese ET, Ryu DT (1980) Enzyme Micob Technol 2:239
120. O'Donnell HJ (1993) MEngSc thesis, University College Dublin
121. Leonard T (1995) MEngSc thesis, University College Dublin
122. Davies PF, Remuzzi A, Gordon EJ, Dewey CF Jr, Gimbrone MA Jr (1986) Proc Natl Acad Sci 83:2114
123. Zhou G, Kresta SM (1998) Chem Eng Sci 53:2099
124. Bronnenmeier R, Märkl A (1982) Biotechnol Bioeng 24:553
125. Wudtke M, Schügerl K (1987) In: Rheologie und mechanische Beanspruchung biologische system. VDI-GVC, Düsseldorf, p 159
126. Baldyga J, Bourne JR, Zimmermann B (1994) Chem Eng Sci 49:1937
127. Baldyga J, Bourne JR, Gholap RV (1995) Chem Eng Sci 50:1877
128. Murtagh J (1994) MEngSc thesis, University College Dublin
129. Ryan M (1997) MEngSc thesis, University College Dublin
130. Hiney J (1999) MEngSc thesis, University College Dublin
131. Panchapakesan NR, Lumley JL (1993) J Fluid Mech 246:197
132. Lasher WC, Taulbee DB (1994) Int J Heat Fluid Flow 15:357
133. Ayazi Shamlou P, Stavrinides S, Titchener-Hooker N, Hoare M (1996) Bioproc Eng 14:237
134. Van Suijdam JC, Metz B (1981) Biotechnol Bioeng 23:111
135. Thomas CR, Jüsten P, Paul GC, Nienow AW (1998) Bioproc Eng 18:7
136. Ayazi Shamlou P, Makagiansar HY, Ison AP, Lilly MD (1994) Chem Eng Sci 49:2621
137. Whorlow RW (1980) Rheological techniques. Ellis Horwood, Chichester
138. Born C, Zhang Z, Al-Rubeai M, Thomas CR (1990) Biotechnol Bioeng 40:1004
139. Thomas CR, Nienow AW, Dunhill P (1979) Biotechnol Bioeng 21:2263

140. Edwards N, Beeton S, Bull AT, Merchuk JC (1989) Appl Microbiol Biotechnol 30:190
141. Soule JC, Wilhelm AM, Riba JP, Ambid C (1987) Entropie 23:24
142. Dunlop EH, Namdev PK (1993) In: Nienow AW (ed) Proc 3rd Intl Conf Bioreactor and Bioprocess Fluid Dynamics. MEP, London, p 447
143. Wilson G (1980) Adv Biochem Eng 16:1
144. Bailey CM, Nicholson H (1989) Biotechnol Bioeng 34:1331
145 Hong YC, Labuza TP, Harlander SK (1989) Biotechnol Prog 5:137
146. Scragg AH, Morris P, Allan EJ, Bond P, Hegarty P, Smart NJ, Fowler MW (1987) In: Webb C, Mavituna F (eds) Plant and animal cells: process possibilities. Ellis Horwood, Chichester, p 77
147. Scragg AH (1990) Enzyme Microb Technol 12:82
148. Leckie F, Scragg AH, Cliffe KC (1991) Enzyme Microb Technol 13:801
149. Scragg AH, Bond PA, Fowler MW (1988) Proc 6th European Conf Mixing. Pavia, Italy, p 457
150. Tanaka H, Semba H, Jitsufuchi T, Harada H (1988) Biotechnol Lett 10:485
151. Ramirez OT, Mutharasan R (1990) Biotechnol Bioeng 36:911
152. Kamen AA, Chavarie C, André G, Archambault J (1992) Chem Eng Sci 47:2375
153. Fukui H, Tanaka H (1995) Plant Cell Tiss Org Cult 41:17
154. King AT, Davey MR, Mulligan BJ, Lowe KC (1990) Biotechnol Lett 1:29
155. Wongsamuth R, Doran PM (1994) Biotechnol Bioeng 44:481
156. Singh G, Curtis WR (1994) In: Shargool PD, Ngo TT (eds) Biotechnological applications of plant cultures. CRC, Boca Raton, p 151
157. Garcia-Briones MA, Brodkey RS, Chalmers JJ (1994) Chem Eng Sci 49:2301
158. Ballica R, Ryu DDY (1994) In: Ryu DDY, Furusaki S (eds) Advances in plant science, studies in plant science 4. Elsevier, Amsterdam, p 221
159. Yang JH, Hu ZD, Wang YP (1995) Chin J Chem Eng 3:101
160. Ducos JP, Pareilleux A (1986) Appl Microbiol Biotechnol 25:101
161. Kim DI, Pedersen H, Chin C-K (1991) Biotechnol Bioeng 38:331
162. Kobayashi Y, Fukui H, Tabata M (1991) Plant Cell Rep 9:496
163. Hegarty PK, Smart NJ, Scragg AH, Fowler MW (1986) J Exp Bot 37:1911
164. Schlatmann JE, Nuutila AM, van Gulik WM, ten Hoopen HJG, Verpoorte R, Heijnen JJ (1993) Biotechnol Bioeng 41:253
165. Smith JM, Davison SW, Payne GF (1990) Biotechnol Bioeng 35:1088
166. Doke N (1983) Physiol Plant Pathol 23:345
167. Jacks TJ, Davidonis GH (1996) Mol Cell Biochem 158:77
168. Levine A, Pennel RI, Alvarez ME, Palmer R, Lamb C (1996) Curr Biol 6:427
169. Delledonne M, Xia Y, Dixon RA, Lamb C (1998) Nature 394:585
170. Apostol I, Heinstein PF, Low PS (1989) Plant Physiol 90:109
171. Doke N, Miura Y, Sanchez LM, Park HJ, Noritake T, Yoshioka H, Kawakita K (1996) Gene 179:45
172. Mehdy MC, Sharma YK, Sathasivan K, Bays NW (1996) Physiol Plant 98:365
173. Lamb C, Dixon RA (1997) Ann Rev Plant Physiol Plant Mol Biol 48:251
174. Low PS, Merida JR (1996) Physiol Plant 96:533
175. Wojtaszek P (1997) Biochem J 322:681
176. Bradley DJ, Kjellbom P, Lamb CJ (1992) Cell 70:21
177. Barz W (1997) University of Münster, personal communication
178. Legendre L, Rueter S, Heinstein PF, Low PS (1993) Plant Physiol 102:233
179. Collén J, Pedersén M (1994) Physiol Plant 92:417
180. Cazalé AC, Rouet-Mayer MA, Barbier-Brygoo H, Mathieu Y, Laurière C (1998) Plant Physiol 116:659
181. Schreck S, Dörnenburg H, Knorr D (1996) Food Biotechnol 10:163
182. Dörnenburg H, Knorr D (1998) Trend Food Sci Technol 9:355
183. Yahraus T, Chandra S, Legendre L, Low PS (1995) Plant Physiol 109:1259
184. Bolwell GP, Wojtaszek P (1997) Physiol Mol Plant Pathol 51:347
185. Mehdy MC (1994) Plant Physiol 105:467

186. Bush DS (1993) Plant Physiol 103:7
187. Knight MR, Campbell AK, Smith SM, Trewavas AJ (1991) Nature 352:524
188. Grabski S, Arnoys E, Busch B, Schindler M (1998) Plant Physiol 116:279
189. Skobeleva OV, Ktitorova IN, Lyalin OO (1996) Russ J Plant Physiol 43:439
190. Ando J, Ohtsuka A, Korenaga R, Kamiya A (1991) Biochem Biophys Res Commun 179:1192
191. Ando J, Ohtsuka A, Katayama Y, Korenaga R, Ishikawa C, Kamiya A (1994) Biorheology 31:57
192. Falke LC, Edwards KL, Pickard BG, Misler S (1988) FEBS 237:141
193. Otte O, Barz W (1996) Planta 200:238
194. Brownleader MD, McNally PE, Davies GEA, Trevan M, Dey PM (1997) Phytochem 46:1
195. Iraki NM, Bressan RA, Hasegawa PM, Carpita NC (1989) Plant Physiol 91:39

Received Juni 1999

Author Index Volume 1–67

Author Index Volumes 1–50 see Volume 50

Subject Index

MIX
Papier aus verantwortungsvollen Quellen
Paper from responsible sources
FSC® C105338
www.fsc.org

If you have any concerns about our products,
you can contact us on
ProductSafety@springernature.com

In case Publisher is established outside the EU,
the EU authorized representative is:
Springer Nature Customer Service Center GmbH
Europaplatz 3, 69115 Heidelberg, Germany

Printed by Libri Plureos GmbH
in Hamburg, Germany